The Insider's Guide to Technical Writing

by Krista Van Laan

The Insider's Guide to Technical Writing

Parts of this book were derived from the book *The Complete Idiot's Guide to Technical Writing*, the rights to which were reverted to the author by the original publisher.

Disclaimer

Trademarks

STC Imprint: The Society for Technical Communication Imprint highlights books that advance the theory and practice of technical communication. For more information about the Society, visit **www.stc.org**.

XML Press
Laguna Hills, California
http://xmlpress.net

First Edition
ISBN: 978-1-937434-03-8
Library of Congress Control Number: 2012935250

Contents

Foreword

Starting as a technical writer in 2012 is so different from starting in 2001, when Krista Van Laan's first book was published. It's even more different since I started in this field in 1977. I'm pleased that Krista has written a new version of her first guide for technical writers. Not only has Krista brought the technology up to date, but she has also stressed what professional technical writers have known for 30 years and more: the importance of knowing your information users and their needs and knowing them better than anyone else in the organization.

Not only should the user be the center of the technical writer's world, but first-rate technical writers must be responsible for understanding the product. Too often, as I'm sure we've all experienced, it seems as if whoever has written the user guide has never actually used the product. That crucial small insight that makes the difference between a successful instruction and one that is confusing and frustrating only comes from direct experience and lots of communication with real users. As Krista emphasizes from the very first, writers need to get out of their cubicles and meet the users.

The Insider's Guide provides exactly the perspective that new technical writers need about teamwork, collaboration, responsibility, curiosity, and more. At the same time, it describes what managers expect today from their writers, even writers with multiple years of experience. The emphasis on flexibility and a willingness to change with the environment is an essential feature of this book.

In Part 1: "Is This the Job for Me?" the advice for people seeking to enter the field is remarkably sound. Education, training, internships, networking, and social media all provide avenues for newcomers to the field. Following the recommendations in Part 2: "Building the Foundation" provides a newcomer with a path to success in finding a job opportunity.

A newcomer on a first assignment is well served by the recommended best practices, especially the focus on knowing the user and knowing the technology. No one should come away from *The Insider's Guide* believing that technical writers are simply formatters of other people's words. In fact, I recommend giving copies to colleagues in engineering, software development, and management who might not understand what the technical writer's role should be. Part

2 makes the best practices of this field clear, demonstrating what good technical writing looks like and how it comes to be.

Part 3: "The Best Laid Plans" begins with the need for planning—from my point of view, an essential recommendation. Too often new writers just start writing without any idea of where they are going or how long it will take. Especially important is Krista's advice to become your own subject matter expert. It's a mistake to think that engineers and software developers will write the content for you. First, they have their own jobs to do. Second, they are unlikely to keep the end users' point of view in mind as they write. That, of course, is your job.

Part 4: "On the Job" focuses on the information a new writer needs about the day-by-day project requirements. I'm particularly pleased with the focus on topic-based writing, especially the Darwin Information Typing Architecture (DITA) standard. Even if you are in an organization that continues to produce books in print or PDF, thinking and writing in topics is an essential perspective. Consider that the topics that you now might compile into a book can easily be transformed in the future into context-sensitive help topics or individual articles on a website, in social media, or on a mobile device.

In Part 5: "The Tech Writer Toolkit," Krista carefully sums up the essentials of the trade. Discussions of style guides, front and back matter, indexes and glossaries, and typography provide a complete toolkit. I'm happy to note the emphasis on writing for translation and understanding the localization process. Writers must be aware of the problems they can cause translators and how to avoid them.

Finally, in Part 6: "I Love My Job…" it's good to hear about the negatives of the field—as a dose of reality.

For many years, experts in the field have collected data that demonstrates the value of sound, usable information on customer satisfaction. We link that satisfaction to improved customer loyalty to a product and a brand. Successful users, happy with products they know how to use, become loyal customers, recommending a product to friends and colleagues as well as investing more for their own use.

JoAnn T. Hackos, Ph.D.
President, Comtech Services

JoAnn Hackos is president of Comtech Services, a content-management and information-design firm based in Denver, Colorado. She is director of the Center for Information-Development Management (CIDM), and a founder of the OASIS DITA (Darwin Information Typing Architecture) Technical Committee and an author of the DITA specification. She is a past president and Fellow of the Society for Technical Communication. Her books include *Introduction to DITA: A User Guide to the Darwin Information Typing Architecture; Information Development: Managing Your Documentation Projects, Portfolio, and People; Content Management for Dynamic Web Delivery; Managing Your Documentation Projects; Standards for Online Communication;* co-author of *User and Task Analysis for Interface Design.*

About this Book

I published my first book about technical writing, *The Complete Idiot's Guide to Technical Writing* (Alpha Books) in 2001 with co-author Catherine Julian. That book was written and published during a very fertile period for technical writers. The "dot-com" bubble was at its biggest. Many high-tech companies that are now household names got their start during that time, and all of those companies discovered that they needed technical writers. A book that explained how to do the job was a great help, not only to newcomers in the field, but also to old-timers who were learning how to do things they'd never done before.

Technical writers today wear even more hats than they did then. Today's tech writers truly are technical communicators, as they build information to be distributed in many forms. A book that explains the big picture is more useful than ever for the tech writer who strives to add value to the company.

As well, the technology has changed so drastically that a book published in 2001 has very little application in the world of today's technical writer. That is why I've written this book for today's world and today's technical writer, the technical writer who's not an idiot, but rather a smart and ambitious person who wants to learn about the profession from someone who knows it from the inside out. While some material is repeated and updated, nearly all of it is new, so even people who read both will find completely new and useful material in *The Insider's Guide to Technical Writing*.

This book is targeted to technical writers at many levels: those of you who are interested in the field and want to learn more about it, and those of you who are just starting out in the field and want to be the best you can be on the job. It also contains much information for experienced technical writers, many of whom would like to know about tools and technology they may not be currently using. If you are a lone technical writer, you are sure to find useful information in this book as you wonder, without colleagues to brainstorm with, how to do something different or better. Finally, I believe this book has value for the beginning Technical Publications manager. Managers need to know about much more than just writing, and this book covers the entire technical writer's toolkit.

I've worked in, managed, and built from the ground up multi-level Technical Publications and User Experience departments in telecommunications, consumer software and hardware, enterprise software and services, and Internet companies. I've used tools you've never heard of and ones you'll use daily, and written every type of content there is as an employee, contractor, and freelancer. I've trained beginners with no more skills than the ability to write in English, who then became highly skilled technical writers. And I've had documentation managers report to me as well, so I'm able to provide information that will help a Tech Pubs manager build and run a better group.

This is the book I wish I'd had when I was starting out and when I was training beginners—a resource that not only tells you how to follow best practices of technical writing (there are lots of those), but also provides specific steps on how to master the non-writing skills that are so important to daily work life—skills like scoping time lines, setting benchmarks, shepherding reviewers, and coping with products still evolving until the day they ship. It's a book that can get you through a major, daunting transition—starting from zero and climbing to a high level of professional competence and confidence. A book about the *job* of being a technical writer.

My experience is largely in high tech, and that environment is the focus of this book, even though technical writing is needed in many other work environments. I believe that if you can succeed in the high-tech world, you can manage anywhere. Although the examples given in this book apply to software, hardware, and website writing, the rules, best practices, and methodology of technical writing can be applied to any field.

Of course, there is still much more that could go into this book. I think of new topics I could have added all the time, and you will, too. You'll find a better way to do something or hear of a different tool or process to use that didn't get mentioned in this book. But I feel confident that with the foundation this book provides, you'll know how to seek more information about the things that interest you, and more importantly, try them out yourself. You won't be in the dark when someone drops a buzzword or mentions an aspect of technical writing that you might not otherwise have heard of.

I'm keeping no secrets here. I've laid out all I know about what it takes to work as a well-rounded and successful technical writer—the kind of writer who gains respect from colleagues all across a company.

If you want a career that lets you play with all kinds of fun technology, interact with smart and creative people, put the keys to high-tech products into the hands of users, and earn a good living by writing, then this book can help you find your way. Whether or not the market is booming, the technical writing profession always has a steady beat!

Krista Van Laan
San Jose, California
April 2012

In This Book

Technical writing is no different from many other careers: you learn your craft, you get started, you build momentum, and then you decide where to go from there. This book follows the same pattern.

- **Part 1. Is This the Job for Me?** orients you in the tech writing field by explaining what the job is all about, the skills you need to succeed, and how you can get into the profession.
- **Part 2. Building the Foundation,** discusses the basics of technical writing and the different types of documentation you'll be expected to deliver. Perhaps the most important consideration is understanding your audience, and a whole chapter is dedicated to helping you do that.
- **Part 3. The Best Laid Plans,** provides a lot of information about the different types of processes you might need to follow, and then helps you drive your own process and learn how to create a schedule. It includes tips on learning the products and technology about which you'll be writing as well as information about the tools you'll be using to produce the documentation.
- **Part 4. On the Job,** is all about doing the job. After reading this, you'll have a good idea about what it's like to walk into a new tech writing job and start being productive right away. Chapters include advice on gathering information, actually writing documentation, and working with reviewers.
- **Part 5. The Tech Writer Toolkit,** contains information about some of the extras for which a tech writer is often responsible. You'll get help on creating a style guide, index, templates, and layouts, and on managing translation and localization work.
- **Part 6. I Love My Job, I Love My Job, I Love My Job...** takes a look at what it's like to be a technical writer. You'll gain some insight into the ups and downs of life as a tech writer. Then I wrap up *The Insider's Guide to Technical Writing* with ideas of where to go from here as you continue and grow in your chosen career.
- **Appendixes** include a glossary to help you understand the terms used in this book and in the field, along with lists of useful books and websites. You'll find plenty to help you continue with your self-education.

Acknowledgments

I'd like to thank all the excellent technical communicators with whom I've worked through the years. As well, thanks to the people who provided their "true stories" for the case studies used in this book. You know who you are.

I also want to acknowledge and thank my parents, who taught me the value of reading, writing, and most importantly, a job well done.

More thanks to the people who helped put this book together:

Editor: Elizabeth Rhein

Cover art and illustrations: Douglas Potter

Author's cover photograph: Jill Holdaway

Layout and design: Krista Van Laan

Sidebars

This book contains plenty of fun and informative extras:

Insiders Know: These are the tips that will help make you a pro. Whether it's an idea on how to do something better or a way to "dodge a bullet," this is insider info you'll be glad to have.

Coffee Break: All work and no play makes technical writers cranky. Sure, you could live without these fun tidbits, but why should you? Everybody needs a break now and then.

True Stories: Real case studies of technical writers, solving real problems.

All contributor's names are pseudonyms.

Part 1. Is This the Job for Me?

You can't answer that question until you know more about what the job involves. This section introduces you to the field of technical writing and shows you what a technical writer looks like. As you read on, you may discover that a tech writer looks like you.

Chapter 1

Calling All Tech Writers

Why tech writing...and why now.

What's in this chapter

- Putting a face to a technical writer
- Where technical writers work
- Becoming a member of a virtual and real community
- Why technical writing can make a life-and-death difference

Every gadget, game, and computer program comes with some form of instructions. One tells us how to use our mobile phone, another how to create a home theater with many different parts, and still another how to use software to fill out our tax forms. And it doesn't stop. New technology comes out all the time, and new products depend on manuals, blogs, and websites to explain them.

Where an aging population meets advances in health care technology, you'll find devices such as blood sugar monitors, cardiac implants, and walk-around chemotherapy, as well as hospital-sized tools such as imaging chambers. All are supported by documentation—documentation that informs, instructs, and saves lives.

It's not only the consumer who has to understand how to use technology. The companies that make consumer products and services also have their own enterprise technology—whether it's databases, statistical analysis systems, or reporting systems—to help run their businesses. The technical folks running those complicated systems aren't born knowing how to use the software that keeps the back office humming. They require documentation to help them learn what to do.

In fact, nobody is born knowing how to use all these tools, devices, and software. (At least not yet!) Like it or not, high-tech gadgets, software programs, and websites aren't always so easy to use. That's why there are technical writers. Until products become so intuitive and simple that you don't need any help running them (or figuring out what's wrong when they don't run), we are likely to *continue* to need tech writers.

Making a Living

We've all seen bad documentation. If you've ever thrown down poorly written instructions, and thought, "I could write better instructions than that!" you are a candidate for a technical writing career.

Certain qualifications give you the best shot at becoming a tech writer: an ability to write clearly and directly, a college education, good organizational skills, and (this is important) an interest in and aptitude for technology and the willingness to learn about the topic you're writing about. You'll find out as you go through this book that there are many skills and qualities, some obvious, some not so obvious, that help to make a successful tech writer.

What Is a Technical Writer?

The United States Department of Labor recognizes Technical Writer as a distinct job category, stating that "technical writers, also called technical communicators, produce instruction manuals and other supporting documents to communicate complex and technical information more easily. They also develop, gather, and disseminate technical information among customers, designers, and manufacturers."

The US Department of Labor Bureau of Labor Statistics website contains excellent information about the technical writer profession, including how to become one, salary statistics, and more. Go to **bls.gov/ooh/Media-and-Communication/Technical-writers.htm**

Of course, the specifics of the job vary widely from company to company, but broadly stated, a good technical writer (you want to be one of those, right?) creates or gathers technical information and then organizes it and presents it in such a way that it is understandable and useful to the defined audience. In this book, I call what a technical writer produces *documentation*. Documentation refers to any content (written, illustrated, or both) supporting the use, operation, maintenance, or design of a product or service. It can be in the form of a printed book (yes, some products still come with printed books), a web page, help in software or on a mobile device, a video, blog, wiki, or more, but it's all still documentation.

What's in a Name?

The Society for Technical Communication, the largest professional organization dedicated to advancing the arts and sciences of technical communication, would call you a "technical communicator." And you can yourself that, too.

Although I use the term "technical writer" in this book, you might go by many different titles throughout your career. Besides "technical writer," you may be called, or call yourself, technical communicator, documentation specialist, or information developer, to name just a few. Sometimes a company creates a new job title to give the writers a career path. Other times, technical writers feel the title of "technical writer" does not encompass all they do, so they assume a title that they feel is more descriptive of what the job really entails.

Although I agree that today's technical writer is truly a technical communicator, and this book will prove that, I use the term "technical writer" or "tech writer" throughout this book because at the time of this writing, that is the term used in job postings, by hiring managers, and throughout the organizations in which I've worked.

And people know what it means. (I've seen the blank stares that can result when someone tells a person not in the field, "I'm a technical communicator.")

> **COFFEE BREAK**
>
> Do you have what it takes to be a good technical writer? Check the boxes and see how many of these characteristics fit you:
>
> - ☐ I love to learn about how things work.
> - ☐ I'm good at giving directions. (Now if only people would follow them.)
> - ☐ I like to teach people and explain how to do things.
> - ☐ I enjoy language and words.
> - ☐ I'm very aware of grammatical errors and typos.
> - ☐ I'm able to work well with many different types of people.
> - ☐ I'm flexible. If something has to change, I don't freak out.
> - ☐ I pay a lot of attention to details.
> - ☐ I'm able to keep track of many things at the same time.
> - ☐ I know tech writing is not meant to be personal expression, so I won't take it badly if someone edits my brilliant prose.

Who Needs Technical Writers, Anyway?

Well, everyone. We rely on tech writers nearly every day. When we put together a piece of furniture from a do-it-yourself store, install a program into a PC or game station, learn how to use our smart phones, use prompts at a self-service kiosk, or read a scientific article, we're reading or using something produced by a technical writer.

Where your own job falls in the corporate organizational chart doesn't always indicate what you will actually be writing. In some companies, the difference between product documentation and marketing or online materials is very clear and rigidly maintained; in others, the lines are fuzzy, if they exist at all.

Technical writers work in a wide range of industries—among them, software, Internet, e-commerce, networking, telecommunications, bioengineering, semiconductor, aerospace, hard science, medicine, automotive, government, heavy equipment, the armed forces, and manufacturing. Although the subject matter is different, processes and methods for producing documentation are often the same across widely different fields. No matter what industry you want to work in, the advice in this book will be helpful.

Moving Fast in a Fast-Moving Environment

You may be surprised by what is required of a tech writer on the job, and it's important that you fully understand what "the job" can entail.

The world where you are very likely to be hired as a technical writer, the world this book focuses on, is the world of high-tech. The environment can be ever-changing—fluid, rather than fixed and predictable—and you may need all the help you can get.

As a tech writer in high-tech, you *must* be flexible, or you will be frustrated by what seems like constant change and shifts in priorities, as companies change plans, or *roadmaps*, to try to predict consumer demand or respond to customers and shareholders. It can seem as if there are no rules or regulations. There's no time to mull over document-creation methodologies or stylistic issues when the marketing folks and the product team seem to be flying by the seat of their pants.

It's OK if you can't handle that type of work environment. You can still be a tech writer in a field that's much more mature, where life moves at a more reasonable pace. And you'll find that the principles and practices described in this book are still very useful.

But if you are up to the challenge, if you want to be part of an exciting industry and make a significant contribution, creating documentation for high-tech companies and their products and services could be the job for you. And believe me, it will be an exciting ride.

Where You'll Fit In

As a tech writer in a high-tech company, you can find yourself reporting into almost any division. A Technical Publications department might be made up of one lone technical writer (yes, that would be you, all by yourself) to dozens of technical writers in a hierarchical structure. Sometimes the larger Tech Pubs departments are centralized, serving the needs of internal users from across the company.

Often, your department will be part of Engineering, which gives you immediate access to the people who design and develop the products. These are the people whose brains you usually have to pick, and the closer you are to them, the better.

You could also work in Customer Support, developing "self-help" content for customers. You might work in User Experience, closely involved with the people who do user research and user-centered design.

> **COFFEE BREAK**
>
> No one becomes famous for being a technical writer, although there are a few famous writers who were technical writers in their past—novelists Amy Tan, Thomas Pynchon, and Kurt Vonnegut, among them.
>
> There are also a few fictional technical writers: Andy Richter played one in his TV sitcom *Andy Richter Controls the Universe*, Michael Harris played on in a 2003 movie called *The Technical Writer*, and Monk's brother is a technical writer on the TV show *Monk*. Perhaps no character has done more—or less!—for our profession than Tina, the technical writer in Scott Adams' comic strip *Dilbert*.

You could work in Product Marketing, Operations, Manufacturing, or Product Management. Or you might work for yourself, at home or at one or more job sites.

Writer or Techie?

Today's successful tech writers usually started out with one of two things: an aptitude for writing or an aptitude for technology. If your aptitude is for writing, be prepared to learn the technical skills that will round out your abilities. Some of today's most successful novelists used to be technical writers, and some technical writers used to be reporters, teachers, or editors. These people, with their talent for writing and their ability to pursue a goal, learned the technology side of the equation and became successful technical writers.

If your aptitude is for technology, you have an important skill to offer employers. Many employers are thrilled when they meet a tech writer with a background in computer or other sciences. When you use the techniques in this book to develop strong writing skills, you will have the makings of an excellent (and very employable) technical writer.

As a tech writer, you may never see your name in print. But you can have a very satisfying career, earn a good salary, help people do things they want to do, and have the pride of saying with absolute truth that you make your living as a professional writer. That last statement is no small accomplishment.

The Accidental Tech Writer

High-tech companies often start out by focusing on engineering and product development and it's not until later, when they have a solid customer base, that they think about hiring technical writers. Until then, the product manager or software developer or quality assurance tester—or maybe no one—writes not only specifications and marketing documents, but also the user documentation.

It sometimes happens that these nonwriters discover they enjoy writing—and a tech writer is born. With practice, books like this one, and training, they can hone their technical writing skills and become among the most sought-after types of tech writers.

More often, these "accidental tech writers" dislike the writing part of the job or find it difficult and time-consuming. They want to spend their time doing what they do best, not writing. And that leaves an opening for people like you.

Making a Difference

Tech writing can be more than just a way to earn money. If your job is to document software, you might be lucky enough to be involved in the excitement of discovery and world-changing technology. (Just imagine how it must have felt to have been a technical writer at a company like GRiD Systems, developers of the first laptop, or NASA in the early days of space travel.)

If your job is to document medical devices, scientific instruments, biochemical or aeronautic software or hardware, or Cirque du Soleil procedures, you can have responsibility for someone's health or life.

Less dramatically, the products or services you end up writing about may touch people's lives in practical, mundane, everyday ways—to make small tasks faster, easier, or more fun. Ultimately, as a technical writer, you make a difference in many ways—to the product, to the company, to the user, and to yourself—and you can have a good time doing it.

Chapter 2

What Does a Technical Writer Do, Anyway?

Typical tech writing tasks and how you'll fit into the product team.

What's in this chapter

- The skills today's technical writer needs to have
- A day in the life—it's much more than you might guess!
- The technical writer as user advocate

When people think of writers, they often imagine someone sitting happily alone in the ivory tower, either lost in thought contemplating a perfect turn of phrase, or typing madly under the influence of the muse of the moment.

It's true that technical writers are writers indeed and sometimes have moments like these. But there's much more to a typical technical writer's day than just writing. In fact, if you're like the technical writers I know, only part of your day will actually be spent writing.

More Than Writing

There is a lot involved in documentation development. Today's technical writer is expected to be proficient in:

- User, task, and experience analysis

- Information design
- Process management
- Information development

Surprised? Those are the basic skills the Society for Technical Communication (STC) expects a technical writer to have to qualify as a Certified Professional in Technical Communication (CPTC). On top of that, you may need knowledge of an array of document development and publishing tools, plus the ability to handle some layout and design, translation management, and quality assurance! Add to that the need for a solid understanding of the workings of your company's products and business needs, and you'll see that writing is just a part of the equation. When you think "tech writer," think "Jack of all trades and master of some."

Certification: which side are you on? Controversy abounds about whether it can matter to a technical writer. Certification allows technical writers, like project managers, networking experts, and others, to show that they bring something extra to the table. Technical writing certification is important, proponents say, because a hiring manager can be sure that a writer has been proven to have certain qualifications.

The anti-certification side contends that the field is so diverse and job requirements vary so widely, no standard certification is meaningful. In addition, certification doesn't prove that a writer has the crucial technical knowledge.

And so the argument goes...and has been going on for years among technical writers. It's up to you to make your own decision about whether this is a path to pursue.

Filling Some Big Shoes

Technical writers are responsible for communicating information that has some specific characteristics:

- It's *what* the reader wants to know—no more, no less.
- It's *where* the reader can find it at the moment it's needed.
- It's *part* of a system of information that all fits together, to mesh with what the user already knows, in a way that causes each component to make sense and be useful.

Sound like a big responsibility? It is! Fortunately, there are tried-and-true ways to create content that meets these needs, and this book will help you learn them.

A Day in the Life

You may greet the workday by learning about the latest emergency, whether it's a new software patch that needs release notes—before lunch—or a last-minute edit in something you're working on, or a request from a colleague to find an old version of a document, or notice that a release will be delayed.

After dealing with the morning's hot issues, you probably will want to start work right away on whatever this morning's top priority is. (And don't be surprised if it's not the same as yesterday's priority or tomorrow's priority.)

Perhaps you are working on a networking optimization guide for your customer's data center and you don't know enough about networking. You feel pressure because the manual is due in four weeks and you haven't started it yet. You spend some time doing research online, set up meetings with several people who you know a lot about the topic, and then search for a course to learn more about what you need to know.

Before lunch, you might attend *standup meetings* for two of the products you're working on. The Engineering division has started using the Agile software development methodology and the writers, as part of the scrum team, attend daily short status meetings in which participants stand in a room and share their status. (Learn more about Agile methodology and scrum teams in Chapter 9, "Process and Planning.")

Just as you're getting ready to go to lunch, the director of Product Marketing asks you to give a quick look at a white paper she's written so she can post it on the corporate website that day. Editing someone else's work calls for tact as well as talent. You eat at your desk while you work on the white paper and are pleased that you caught a couple of potentially embarrassing errors while improving the organization and style.

White paper is an industry term for a document (like a report) that states a position or helps to solve a problem. White papers are used to educate readers and help people make business decisions. A white paper can be very technical but it must also be clear and easy to read. Writers who can write white papers with their balance of marketing and technical-speak can often do very well.

Later, you're asked to attend a meeting to talk about what documentation will be needed for a new product's upcoming release. Because the company can't release the product without documentation, your contribution is an essential part of the customer delivery. Along with the standard help and manuals, Product Management is asking you to provide content for a blog and a Twitter feed and for a YouTube video that will be essential parts of a marketing strategy to appeal to a broader audience.

When you are finally able to sit in front of your computer for a couple of hours of uninterrupted writing work, you close your email and put on your headset with music to drown out the office sounds. (A tech writer, although typically part of a larger team, frequently works independently, spending long periods of time alone creating content.) You are writing content for the internal support site and you promised to have it finished by the end of the day. Some of this content will be tagged for the external customers' knowledge base. It's important to pay attention to what may be released to the outside world and what must remain internal.

Not reading email for two hours means that there are many new messages by the time you take a break. Several of them contain review comments for the draft you sent out to subject matter experts a couple of days ago. Each reviewer has a different method of giving feedback—one marked up a paper copy with a red pen and left it on your chair while you were away from your desk, others typed all their comments in email, and still others used the text-editing tools in the PDF review version you sent.

INSIDERS KNOW

SME: Pronounced "smee," by some and "Ess em ee" by others, this is a common tech writer acronym for Subject Matter Expert.

The SME can be anyone from the inventor of the technology to the guy in the mail room—it all depends on the subject matter expertise you need. But behind that term is a real person—a person you will depend on.

Some reviewers require good old-fashioned face-to-face contact. You rush to catch one of the subject matter experts before he leaves for the day and it turns into an hour-long meeting to understand some of the issues brought up during the review. As you listen and take notes, you ask questions to keep the conversation headed in the right direction. It's your job to make sure all of the content is correct, no matter how technical and specialized it is.

The end of your day finds you sending out a couple of last important email messages. Some of the development team is in a time zone 11 hours later and you need to ask some questions that only they can answer. If you email them now, their answers may be waiting for you in the morning. Other emails go out to members of the Tech Pubs team—you were so busy, you didn't get a chance to meet with two of your teammates before they left for the day, and the three of you are working on documentation for the same product family. You need to sync up some information with the two of them before you can move forward.

Before you go home, your boss stops you and asks if you could send your current draft to the account representative to be given to a potential customer. Darn! You thought you could leave at a reasonable time tonight. Well, you know how important it will be to win this particular customer, and you're excited that your documentation could help make the sale. You head back to your desk to log in again and send out the PDF.

Sound like a busy day? It is. And there are still many other areas of your job you didn't get to today—proofreading, creating online help, assisting an engineer with a presentation, working on templates. Oh, yes, and there's the document plan you promised your boss you'd have by the end of the week. Now you see why many tech writers want to be called something that reflects the many hats they wear—technical communicator or information developer or anything other than plain old technical writer!

Turning "Geek Speak" into Plain English

Talking with that software developer earlier, you were reminded of why you are important to the company. Engineers, developers, and other technical specialists often have one thing in common: their high level of expertise makes it difficult for them to think and communicate at a level all users understand.

Engineers who are asked to write for customers are so familiar with their own technology, they often don't realize they are leaving out crucial information. They make assumptions based on their knowledge and assume the reader has made them too. That can make it difficult or even impossible for a typical user to follow their train of thought.

Not to mention that they'd like to be left alone to do their jobs, thank you, and their job is not writing product documentation. It's up to you, the technical writer, to bridge that gap—or yawning chasm—between what the expert knows and what your target audience needs to know. Some of your job is to act as a translator, sorting out relevant content and presenting ideas in ways that make sense to readers who don't "speak geek."

Figuring Out What Comes First (and Putting It There)

Today we are all swimming in information overload. As a technical writer, a major part of your job is to organize information. From everything you *could* say, you must figure out what you *should* say, how you say it, and where it belongs.

Begin at the beginning and figure out what the starting point is. What has to be done before anything else can happen? Does one piece of hardware have to be connected to another? What kind of cable should be used? Are there drivers that have to be installed? Does an account need to be set up? Do you have to unzip a file onto a server and rename it something else? What about the user interface—which parts of the product does a user need to see first to perform critical business tasks?

Organizing ideas often comes easily to people who can write, and with practice it becomes easier. You'll be surprised how much the other members of the teams you work with depend on you to supply the logical, fundamental starting points in discussions and meetings as well as in written content.

Writing and Maintaining Documentation

Whether what you write is ultimately delivered by clicking a link on a browser, by downloading from an extranet, viewed on a phone, or printed on paper and included in a box, there are many deliverables for which you'll be responsible, all of which become part of your organization's product.

Often you'll write brand-new content where none existed before. It's exciting to create something out of nothing. But as products continue to grow, evolve, and mature, the documentation has to as well. New product features and functions are added, and the documentation for the product—like online help, knowledge base content, and the user guide—has to be updated.

Maintaining, updating, and adding to documentation through the lifetime of the product will be a major part of your job. This can mean adding new information as new features are incorporated, or removing obsolete information. It may also mean revising and improving the writing—whether the original writer was you or someone else, there may not have been time to do the best job on the first round.

You might also learn that there were errors or inaccuracies in the earlier version and they need to be corrected in this one. It's up to the technical writer to make sure the body of information about a product doesn't become a work of fiction.

Understanding How Things Work

Writers generally are expected to have mastery of the written word, but in technical writing that's not all you need to master. Understanding your company's products, their basic functions, and how they differ from each other is a key part of your job. You also need a deep understanding of the people who buy and use your company's products—the audience you write for.

You'll often be asked to act as the in-house "explainer" between one department and another or as a resource for newly engaged employees, consultants, or contractors. With your product knowledge, you could even be asked to step in as a trainer to customers or other employees.

Ideally, you'll want to be assigned to one product line or family until you become very familiar with it. Then you'll be able to write about it yourself without being simply a transcriber, and you'll become a valuable part of the team.

It's unlikely that anybody expects you to understand everything at the same level as an engineer, developer, or product manager. You are, however, expected to learn enough about the product and the product family so that you can write about the product and make informed decisions about what content needs to be made available to the customer.

Being a Catalyst for Change

An unexpected aspect of being a technical writer is that it puts you in a position (sometimes against your will) of "stirring the pot" and initiating changes in product appearance, product function, and sometimes even how and to whom a product is marketed. How does this happen?

Well, it's a funny thing, but when a technical writer starts asking questions about a product, people start looking at it and thinking about it in ways they didn't before. And when a technical writer brings people together from different departments who might not normally talk with each other, there can be constructive dialog that otherwise would not have occurred. It can be very exciting to sit back at a lively meeting and realize you provided the spark that ignited all this exchange of ideas. One of a technical writer's most important functions is to spark discovery—sometimes in the most unlikely places.

Sometimes it's as simple as giving everyone their first good, clear look at what a product really does or how it really acts. Perhaps nobody realized that it takes nearly 30 steps to configure and install the software until your documentation spelled it out for them in black and white.

The Technical Writer Is the First End User

It's not uncommon for product designers and developers to get so caught up in how to make a product *do* something, that they forget why that functionality is needed. They can lose sight of the needs of the real person who may be depending on that product to perform a task fundamental to her job's success, safety, or convenience.

Here's where the technical writer comes in. Because the user is the very person you're writing for, your job is to ask questions and make judgments from his point of view. The nature of the questions you need answered makes you the first user, before the product is ever released. And it also can put you in the position of being the user's advocate.

This doesn't mean that you should be ignorant of the product that you are documenting. It's a mistake to think that your ignorance gives you a new-user perspective that will let you create better documentation. No, all that does is let you write documentation that adds nothing to what new users can figure out for themselves.

As the user advocate, you need to be able to understand what the user needs while still possessing, and imparting, a higher-level knowledge that helps that user. I talk about this a lot throughout this book, because it's that combination of knowledge and user perspective that will help you become a really good technical writer.

It is how well a product or service ultimately meshes with the needs and wants of the customer that determines the success of the product or service—not only its commercial success, but also its success in serving the user. Too many companies have failed to understand what the members of their target market really need, want, and will pay for. You, by putting yourself in the shoes of the customer, can provide a vital perspective.

Chapter 3

Having the Write Stuff

Certain qualities make a good technical writer. Do you have them?

What's in this chapter

- The qualities that make a good tech writer (Yes, writing is one of them)
- Flexibility, the key to success
- Juggling as part of your job description
- How to say "no" while making them think you said "yes"
- If the big shoes fit, wear them

Now that you have a better idea of what technical writers do, and maybe you're asking yourself whether you have the right stuff—or "write" stuff—to succeed as a technical writer in the fast-paced world of high tech. Or any world at all, for that matter.

There is probably no such thing as the perfect technical writer—everyone has unique strengths and weaknesses—but some attributes can be worth their weight in gold, while others are a bit more lead-like. In this chapter, let's take a look at how important (or not) writing skill actually is, and clue you in on all the other essential skills, including ones you probably never thought of.

But Can You Write?

Many people—or their bosses—think that anyone can write. They all learned in grade school, they say. Those people also sometimes believe that if someone knows a lot about a particular technical topic, that person should be able to produce excellent technical documentation. After all, knowledge is the most important thing, right?

Well, not necessarily. You do need to be a good writer, and not everyone has developed that skill. This means writing clear prose without extraneous words. And while it's true you don't need to be able to write a Shakespearean sonnet to be a good technical writer, you do need to be familiar and comfortable with the fundamentals of writing, especially with the best practices of technical writing discussed in Chapter 6, "Best Practices Make Perfect."

Let's assume you have basic writing skills, you *like* to write, and you are ready to follow all the technical writing best practices covered in the rest of this book. There are still other characteristics and skill sets that will help you succeed on the job.

A Natural Curiosity

If you enjoy technology and finding out how things work, you have the aptitude for being a good technical writer. You may enjoy having the latest version of the latest new gizmo, but tend not to even glance at the user manual that came with the gizmo.

Many tech writers don't read manuals themselves when they start using a new product, service, device, or app—they just dive right in and expect to figure out how it works. Why? It's not out of disrespect for their profession—it's because they like to investigate, explore each feature, and figure out what it does. When they do need the manual, however, you'd better believe they expect it to have exactly the information they want!

Being Technically Adept

Technical capability is important. After all, there's a reason the word "technical" is part of your title. While it's important to be able to interview developers and other experts to gather and write information, a tech writer should not expect to simply collect information, correct grammar, format it, and be done. A technical writer is not the same as a reporter and is definitely not just a "scribe."

A technical writer should have the ability to learn enough about the subject matter to write basic documentation with little or no help and without extensive reviews. If you document software that runs on Linux servers, learn enough about Linux commands to understand what you are seeing and to avoid mistakes in the commands. If you are documenting medical equipment, learn how it works. If you are documenting an end user product, learn what the user needs to do. Chapter 10, "Become Your Own Subject Matter Expert" provides some tips on how to learn about the products you are documenting.

Staying Flexible

If I had to choose one characteristic as the most important a technical writer could have in today's often-stressful corporate life, it would be flexibility. If you're not able to roll with the punches, you're not going to be able to survive.

Tech writers often work up to the very end of the release—or even afterwards—dealing with continual last-minute changes, insufficient information, and little thanks at the end. Tech writers can feel as if they have no control over their own destiny as they are asked to change course numerous times. Release dates can move again and again. Priorities shift, and often technical writers are the last to know.

An experienced technical writer does not break a sweat when asked to change her focus from one "emergency" to another, and then back to the first one. The ability to course-correct, keep a cool and cheerful demeanor, and act as if the new thing you are expected to do now is the thing you always wanted to do are very useful traits for a technical writer.

In the 1960s, doctors discovered while working with patients with severe epileptic seizures that each hemisphere of the brain processes information differently. The left hemisphere is dominant in verbal, analytic, abstract, and logical activities. The right hemisphere is dominant in nonverbal, analogic, intuitive, and spatial activities.

Which side do you think is dominant in tech writers? Some people become technical writers because they are logical, analytical people who like to work with facts and orderly data. But a writer also has to dig into the right side of his brain when it's necessary to approach an idea visually, to design a document for maximum effectiveness, and especially to communicate with the many different types of people that come together.

Attention to Detail

A good technical writer has a natural ability to follow up on details. You're the one who finds the one typo in written material or notices television personalities' misuse of words. You remember what promises were made, follow through on commitments, and keep track of important dates.

These qualities will help you produce documentation of exceptional quality on the job. You won't let your documentation go out with missing information or typos. If a paragraph doesn't adhere to the style guide, you'll feel compelled to fix it to make it consistent. If a developer tells you he'll be able to speak to you "next week," you'll be at his desk first thing Monday morning.

Juggling Flaming Sticks

During your tech writing career, you will often feel as if you are holding many things in the air, all of them emergencies, and all ready to explode if dropped. You must try not to panic as you juggle projects, priorities, and demands on your time. A good tech writer has time management instincts and the ability to multitask. It's not uncommon to be working on many projects simultaneously, all at different stages in their development, each with its own deadline. Aside from keeping the projects themselves straight, you keep track of where you are in each one, and meet the individual deadlines.

A good technical writer is equal parts project manager and writer, with a bit of reporter and private detective thrown in for good measure.

Beef up your project management skills with a class or a book on the subject. Project management is a learnable skill and the Project Management Institute (PMI) administers the Project Management Professional (PMP) credential, an industry-recognized certification for project managers. Having PMP certification can help you get a job as a technical writer; it's a credential hiring managers recognize.

Check it out at **pmi.org**.

You are likely to work with completely different teams, each of which has its own requirements and expectations, and sometimes all those dates change at the same time. I did say flaming sticks, didn't I?

While you don't have to be a certified Project Management Professional (PMP), it certainly helps to have some project management skills as you balance multiple projects against tight deadlines.

If you can accept that juggling flaming sticks is still just juggling, and if done right, is no more daunting or dangerous than with rubber balls, you have the makings of a good tech writer.

The trick, of course, is in knowing the right thing to grab, knowing what to do with it, and deciding when to let go. Chapter 11, "You Want it *When?*" gives you strategies for making those crucial judgments every day at work.

"Get it?" "Got it!"

Quick—what characteristic of high-tech industries makes them unlike all other industries in history? No, it's not the silicon chips. It's the pace of work and development, which seems to match the ever-increasing speed of microprocessors and communications networks. Never before has so much activity been compressed into such small increments of time.

Changes that took other industries years or even decades to experience now happen in only months, or even days. A new technology is introduced one day, and by next week it seems to be everywhere. Something that happened six months ago is essentially ancient history.

Because of the fast pace of change, one of your keys to succeeding as a technical writer is your ability to "hit the ground running." This means being able to grasp the essentials of your company's product line or service quickly and thoroughly as well as understanding the value the customer finds in whatever your company is marketing. It also means being able to respond very quickly to the needs of everyone else in your company. You provide a service not only to the external customer, but also to your internal customers.

The rapid pace of work means your time line for planning, writing, and finishing content has to be carefully managed. There is no time to wait for the muse of inspiration to strike when that countdown clock is running.

Getting Along

Whether you are part of a centralized Technical Publications department in a giant corporation or the lone technical writer at a small startup, you'll spend a lot of your time working with people whose jobs and communication styles are very different from yours. What does this mean for you?

Several things. The technical writer often depends on others to provide critical information during intensely busy periods. It's important for you to be able to both get along with and communicate with your colleagues, whatever that takes, from the most extroverted salesperson to the shyest engineer. More than most, you will be working in what are called *cross-functional teams*, consisting of individuals from more than one organizational unit or function.

Good technical writers are expected to work with everyone at the level necessary for successful communication and collaboration. Sometimes that means developing a keen sense of communication styles—you have to realize that one person talks a *lot* when you're trying to get information, and another person's abruptness isn't anything to take personally, and a third person is in a bad mood on Mondays.

The tech writer is often at the center of a project, performing a service for people from all divisions of the company. A technical writer is frequently expected to make things happen without having any actual authority over the people who are supposed to do those things.

> **COFFEE BREAK**
>
> What's your personality type? The Myers-Briggs Type Indicator uses four different categories: Extroversion/Introversion, Intuition/Sensing, Thinking/Feeling, and Judging/Perceiving.
>
> INTJs make up only two to four percent of the population, but many technical people, including technical writers, are in this group. Knowing what types your colleagues are (and what you are) can help you communicate more easily with others.
>
> Learn more about Myers-Briggs and the similar Keirsey Temperament Sorter at **myersbriggs.org** and **keirsey.com**.

As well, technical writers work with other writers, often working on different documentation deliverables for the same product family. Sometimes two or even more writers work on a single document. It takes a good team player to work well with others on the same project. It can mean compromise as you come to agreements on structure, terminology, and milestones. It may mean agreeing to a schedule and standards driven by someone else. And it means regularly referring back to the work of others to make sure it all syncs up.

You also need to be sensitive to your own manager and her expectations of you. Sometimes you may forget you have a boss, because you answer to so many different people and work so independently. Nonetheless, it's important for you to keep your manager informed of what you are doing at all times. Your manager may expect you to work independently, but she is still responsible for your assignments and performance, so she needs to be kept in the loop.

I'll bet you didn't know a technical writer had to be such a diplomat, did you?

Saying "Yes"

Saying "yes" is a lot of fun and something every tech writer is expected to do...often. After all, Technical Publications provides a service to the rest of the company, and technical writers are often service-oriented types of people.

As a tech writer you'll have to say "yes" more than you'd like to and sometimes more than you ought to. Yes, you need to be flexible and willing to do a lot of things at the last minute; sometimes more things than you thought that "last minute" could accommodate. You also need to know when to draw the line.

When you do say "yes," be sure to be clear about what you are saying "yes" to. People love to hear it and might take it to mean you'll fulfill their hearts' desires. If someone wants the moon, make sure he understands that you're agreeing to an orbiting satellite, not a planetary body. Don't promise what you can't deliver.

Be sure to follow up your "yes" with an email that spells out exactly what you agreed to—including, if possible, specifics such as the date the project is due and a description of what you will provide. This can go a long way toward avoiding later conversations that start with, "But I thought you said you were going to..."

Saying "No"

There is no doubt that saying "yes" is much nicer, but sometimes you just have to say "no." This can be trickier than it sounds. It's nice to make people happy but you could find that sometimes your "yes" got you into big trouble as you realize you might not be able to deliver. If you ignore the deadline, it can cause ripples in all directions—often with disastrous results. You do not want to be at the epicenter of those ripples. They can have a way of turning into tidal waves.

Maybe you don't have the experience to realize how much more you can (or can't) fit into an already tight schedule. If you aren't sure, ask your manager or someone with more experience. You may feel really good when you say "yes" to yet another request, but if it's too much, one of two things will happen: either you'll work all night to finish the job, or you'll jeopardize the delivery.

It's much better to tell the truth, stated in a tactful way, of course, than to let people expect something that can't be done. The best way to say "no" is to say it straight out but follow it immediately with what you *can* do. For example, when asked if you can provide a customized document within the week when you already have several major deadlines, you can say, "I can't give it to you by Friday. But I could get it to you by Wednesday of next week."

INSIDERS KNOW

If you are asked to accept a project from someone other than your manager and you're not sure how it fits in the product roadmap, tell that person you're happy to work on it but you need to run it past your manager. Remember, if you make your manager look good (or bad), she will remember that later on, when it comes time for pay raises and promotions.

You won't always be able to respond with a definite answer, so be ready for those times, too. Have a reply in mind such as, "My manager has assigned me a number of projects that he says take top priority, but I'm sure I'll have some time after those are complete. Shall I call a meeting with him?" It's essential to convey your "no" clearly but without slamming the door in the other person's face. The work will most likely have to be done at some time, so it's best to maintain a good relationship with the requester.

Being Part of the Business

Being a team player is more than just being able to get along with other members of your department and division, and more than being able to work jointly on deliverables with other team members. It also means that you are an employee of the company, and you understand and do what is best for the company. It means that you are aware of the business and its priorities. You know about all the products—not just the ones you work on—and you know who the important customers are and what can help the company be profitable.

Some technical writers feel that they do not have to participate in day-to-day activities and instead can work at home on discrete projects while staying out of the daily grind. When important rush jobs or meetings come up, the writers who are off the premises are not available to share in the work. These same tech writers then feel that they are not treated as part of the team, or that their work is dismissed.

Take a look around you and see what co-workers do. And that means all of your co-workers in other departments such as Engineering and Product Management, not solely in the Technical Publications department. Do people come into the office every day? Then you should expect to, too. Chapter 22, "Working Outside the Box," contains some ideas for staying involved if you *do* work off site. But it's about more than where you do your actual writing; being part of the business means understanding, and participating in, the big picture.

Being Dependable

With schedules so important and *time to market* (the time from inception to release) the factor that can make or break the company's bottom line, reliability is a winning quality in today's working world. Colleagues appreciate it when you return email messages and phone calls. Your boss appreciates it when she asks you to do something and it gets done when you say it will be. The project managers appreciate it when you anticipate problems and can tell them ahead of time if your delivery is at risk.

Some people think that writers don't always pay attention to deadlines because they are more interested in turning a beautiful phrase than noticing the calendar. If this is you, it's something to work on. Dependability is noticed and greatly appreciated in today's working world, simply because a remarkable number of people don't exhibit it.

Everybody appreciates it when you meet the deadlines. Even more than writing skills or the number of airborne flaming sticks you can launch, dependability is one of the best assets you can have. It's doing what you say you'll do, when you say you'll do it.

Chapter 4

Breaking Into the Field

Overcome challenges to land that first tech writing job.

What's in this chapter

- The kind of background you need
- How to kick-start your resume
- Getting that crucial first interview
- Networking tips
- Be a bulldog, but a nice one

So you've decided you've got the "write stuff" and you'd like to get into the technical communications field.

But you have no clue how to go about getting your first job in the field. Perhaps you have little or no writing experience and your college degree is in something like Ancient Civilizations, or you've been working for years in a field not commonly associated with technical writing. You've heard about all kinds of people who have less experience than you who have somehow stumbled into the field and are now making good money as technical writers.

There are plenty of technical writers who find it easy to get hired. Their resumes contain keywords like *pharmaceutical* or *networking* or *Cisco certification*. They have worked for companies we've heard of. Their resumes show that they know about the products they write about. They have both technical knowledge and writing experience.

How can you become one of those people?

This chapter gives some guidance on how best to prepare yourself for a career in technical writing, and then how to impress a potential employer. It takes some work, but this work uses the same skills you'll be employing as a tech writer: gathering information, interviewing, and following up on leads.

All Roads Can Lead to Tech Writing

Luckily for you, successful tech writers come from many different walks of life. There's no agreement about what the best background is. Some attended school specifically for technical communications, but most did not; some have technical educations or backgrounds, but most do not.

The good news out of this mixed bag is that if you have nothing more than the desire, the motivation, and the ability, you can add a few ingredients, and put the pieces together to become a technical writer. There's no secret society ritual, no magic medal conferred by the Wizard of Oz that suddenly makes you a technical writer. What makes you one is the ability to do the job well. What makes you an *employed* technical writer is the ability to convince an employer that you can do the job.

To become a technical writer in today's world, you need to focus on four areas:

- **Education.** If you don't have the right background, there are courses you can take to bolster the education you have.
- **Tools knowledge.** An important part of the tech writer arsenal is mastering the tools tech writers use. There are ways to do it without spending a fortune.
- **Experience.** If you don't have the right job experience, find ways to get it and then make sure your resume, as well as your online presence, reflect that experience.
- **Networking.** Studies show that most people find jobs through networking. A would-be technical writer has many ways to network successfully. You should take advantage of them.

Degree or Not Degree? That Is the Question

Yes, having a college degree (in any field) definitely will make it easier to get a job as a technical writer. Maybe you're considering returning to school to get one and wonder if the degree should be in technical communications. Do you really need it?

The short answer is no, you don't need to have a degree specifically in technical communications. The long answer is that a college degree *is* important. You will meet plenty of technical writers who have no college degree, but they also typically have years of experience.

A degree in technical communications has its advantages, but is by no means essential for becoming a technical writer. If you already have a degree in any field, there's probably no need to go back for another one. Technical writers have degrees in subjects widely ranging from art to zoology, as well as the more likely technical communications, journalism, and computer science. The point of *any* degree is that it shows that the tech writer knows how to learn, and how to stick with something—two things every tech writer must do every day.

If you have not yet decided what to major in or what to study, the right education can be a big help in getting you into the field. Many colleges and universities offer degrees in technical communications, and if you know you want to be a technical communicator, there's no better way to announce your intentions than to major in the field.

You are less likely to find a job in the field with a degree in English, creative writing, or a similar liberal arts subject. However, you can increase your chances of being hired by focusing on the technical aspects of your education. A graduate with a double major in computer science and technical writing has a much better chance of being hired than a graduate with just an English degree or just a technical communications degree. A major in computer science or engineering with elective courses in technical writing, or any kind of writing, also give you a great advantage. After all, there are two parts to technical writer: "technical" and "writer." Now is a good time to put those two parts together.

College Internships

One of the best ways to gain experience is to become an intern. Many companies have excellent intern programs, where they hire students for the summer, pay them, and even give them some benefits. Usually the plan is to hire the intern when she finishes school, but even if that does not happen, it becomes an important job to put on a resume. If you are a student, keep your eyes open for these opportunities.

But the requirements even for interns are steep. Let's look at some excerpts from recent technical writer intern job postings:

> Fundamental engineering skills
>
> Technical expertise in Linux, system administration, programming, or databases

Proven ability to document Java APIs and Java programming tasks

Solid knowledge of SEO (search engine optimization) principles

In-depth knowledge and understanding of social media platforms and their respective participants

Must be a college junior or senior enrolled in any computer-oriented degree program (CIS, Computer Science, etc.)

It can be a bit discouraging if your only technical experience is in creating your own Facebook page. Employers want technical writers with technical expertise.

Mix It Up

If you have a degree and it is not in technical communications, additional coursework in technical writing is not a bad idea. There are many good programs out there, both online and brick-and-mortar. The non-degree technical writing programs give you an overview of technical writing, teach you how to use some of the tools, and help you assemble a portfolio for job-hunting.

Many schools offering these programs use professional technical writers as instructors. These working instructors give you insights about on-the-job reality, not textbook theory, and can be valuable contacts. Your fellow students, too, will be good contacts for the future. (Yes, this is networking, which will be discussed in more detail later in this chapter.)

What the Employer Wants Is What You Want

Check out the job postings in your area and see what kind of experience companies are expecting their technical writing candidates to have. You are likely to find requirements for an array of tools and technologies:

- **Building books** with Adobe FrameMaker, Microsoft Word, and Adobe Acrobat publishing software
- **Reusing content** with XML and text editing tools such as PTC Arbortext or XMetal
- **Creating graphics** with Adobe Photoshop or TechSmith Snagit image-editing software
- **Developing help for software, the Web, and mobile devices,** with MadCap Flare, Adobe RoboHelp, or WebWorks ePublisher help authoring tools
- **Writing for the Web** with Adobe Dreamweaver or other HTML-editors
- **Designing flowcharts and technical illustrations** with Microsoft Visio and Adobe Illustrator programs

- **Sharing content** with Microsoft SharePoint, a wiki, or a content management system
- **Developing interactive learning and tutorials** with TechSmith Camtasia or Adobe Captivate

Although many tech writers will tell you that it's the writing ability that counts, not what desktop publishing or help authoring tool you know, hiring managers often feel differently—especially when there are a lot of applicants for the same job and some of those applicants already know the tools the company uses.

Go ahead and acquire the tool skills that local employers are looking for, either by taking a class or getting the software and learning at home. Nobody cares where you learned the skills; it's just important that you know them well enough to do the job.

Save While You Learn

It can be prohibitive to buy all the software you want to learn, but there are ways you can do it without paying top price.

- If you're a student, take advantage of your student discount. Educational discounts can be huge and give you the opportunity to upgrade later. Upgrade software is much less expensive than buying the full installation software.
- If you're not a student, you'll be happy to learn that many, if not all, of the products you need to learn are available from their company websites on a free trial basis. On your own or with one of the many teach-yourself books, you'll be able to learn them well enough to be productive on the job.
- Download open-source versions of the alternatives to the standard brands of tools. *Open source* refers to source code shared with and among developers and users, resulting in free or low-cost development. You can find software for image editing, layout, diagramming, illustrating, screen captures, and more. If an employer asks you if you know a certain brand-name tool, you can reply, "I'm proficient in the open-source version, which is very similar."

Many of the tools discussed in this book have open source equivalents. You can save yourself or your company big bucks by using some of them. Just do a search online for "open source alternative" plus the tool of your choice or go straight to **osalt.com**, the "open source as alternative" site.

You can also buy used software, but use common sense if you go this route. Make sure that what you buy is a fully functioning version of the software and that you receive the license and serial numbers with your product.

Chapter 12, "You Want it *How?*" talks about the types of tools you might use to develop different types of documentation.

It Helps to Speak the Language

Besides knowing how to use the tools, there's the all-important matter of knowing how to work with the content you create. With so much user content on websites, it is important to know HTML, a markup language that is the publishing language of the World Wide Web. Skills in JavaScript, CSS (the style sheet language that enables you to consistently format a site), and XML (A metalanguage that allows users to define their own markup tags) greatly increase your ability to create content for any company in any line of business.

There are also programming languages, such as C++ or Java. While you probably won't use these to create documentation, familiarity with programming languages is very useful. If you work in the software field, the ability to read code will be especially advantageous.

But First, Are You Experienced?

OK, so you're working on laying down your educational foundation, including learning some of the most important tools. Whether you are a college student, a new graduate, or a career-changer, you might feel as if you're facing an insurmountable challenge: with your current qualifications, no hiring manager will respond to your resume. It's the old Catch-22: you can't get the job without experience, but you can't get experience without a job. What's an aspiring technical writer to do?

When creating samples for your portfolio, it's always better to present something that has actually been used or looks as if it were real. A hiring manager at a company like Cisco or Oracle is unlikely to be impressed by sample instructions for making a peanut-butter sandwich. Find an actual product and write usable documentation for it.

Without question, you will need to build your portfolio. And as you do, you will rewrite your resume to showcase the experience you do have. Your resume should be a living, breathing, continual work in progress.

Build a Portfolio

Your portfolio is very important. Unlike many jobs, where you go to an interview empty-handed and fumble through questions about where you see yourself in five years, technical writers are typically expected to display their skills and abilities by showing the work they've done and talking about the process they went through to do it.

If you have no writing samples, make some! There are several ways you can build your portfolio, although, like most of the suggestions in this book, they all require work. (Groan if you must, but this is work that will pay off.)

As a would-be technical writer who is struggling to break into the field, you may even have to be ready to do some work for free to enhance your portfolio:

- **Volunteer to work for an organization that can benefit from technical writing.** Plenty of nonprofit organizations need help not only with their newsletters, but also with their policies and procedures and software programs. Do a web search on "Technical writer volunteer" or check a site such as **volunteermatch.org**. Talk to nonprofits in your area that are doing work that interests you and see if there are any opportunities for you.

- **Volunteer to write grant proposals for some of these same nonprofits.** Proposal-writing is its own specialty, and many people make a living at it. **grantwritersonline.com** and similar websites offer advice on how to write proposals. There are also proposal-writing classes you can pay for if you want a structured learning environment.

  The best thing about writing a successful proposal is that you can state on your resume that your writing helped bring funding to a deserving organization. Do a web search for "volunteer grant writing" or "volunteer proposal writer" to make a connection.

- **Write product documentation for a sideline business or a startup.** Find someone who owns or is a partner in a small startup and think of ways you

> **TRUE STORIES**
>
> I know three partners who have started their own business building software for financial companies. These partners were grateful when a tech-writer-in-training volunteered to help with their documentation. The partners are happy because their product looks more professional and can attract more customers. The writer is happy because he now has genuine work product for his portfolio and resume.

can contribute. New business owners with little money are grateful for help with product documentation.

- **Write documentation for open source, freeware, or shareware software.** Yes, the same source code I suggested you use to gain tools knowledge for free can also help you build your portfolio. The FLOSS Manuals foundation at **flossmanuals.org** helps develop free manuals for free software. Also try websites like **opensource.org**, **sourceforge.net**, **openoffice.org**, **ifixit.com**, and **opensourcewindows.org** and see how you can help.
- **Write or rewrite documentation for a product that currently exists.** If you've tried and failed to find volunteer opportunities, create samples by writing documentation for products you have. For example, if you are eager to work in the field of multimedia, develop a user guide and help for a media player you own.

You would be surprised what some Tech Pubs managers do to filter out applicants. They can be very critical of technical writers who don't use professional formatting techniques on their resume.

Most resumes in the United States are done in Microsoft Word and experienced Word users know to use styles to format their resumes. That means avoiding pressing **Enter** to add extra line spaces; instead, create a style that adds extra space to the line. Don't use the Normal style in your resume; instead, create a unique style that always looks the same on any computer. Use a template or get help from a Word expert if you aren't familiar with styles.

Recast your Resume

They say a hiring manager spends 10 seconds looking at a resume before deciding whether to give it a more thorough read, so you don't have long to make a good impression.

Make sure your resume is written in a way that casts you as a technical writer. That means emphasizing every bit of writing experience you can lay claim to. If you had a position that wasn't called "Technical Writer," but technical writing was actually one of your job responsibilities, go ahead and put "Technical Writer" in parentheses after your job title.

An obvious tactic that nevertheless eludes some writers is to write your resume as a technical writer should: effectively, succinctly, and correctly. It should show off your writing and layout abilities.

There are many ways to create a resume and not a lot of hard and fast rules about what format, page length, or style works best. However, there are a few things you should remember to do for the technical writer resume:

- Include a summary of qualifications at the top that says you are a technical writer.
- Make sure the format and layout are attractive and professional-looking.
- Make sure all those tools you learned are featured prominently where the hiring manager can't miss them.
- Run your spell-checker. I'm always surprised at how many technical writing job-seekers send out resumes that contain typos. Hiring managers won't overlook that.
- Read and reread for accuracy. Then have an experienced technical writer or Tech Pubs manager read it again.

Making the Most of What You Have: For the Career Changer

Before you decide to bail out of your current job, see if you can turn it into a technical writing job. If your company has a Technical Publications department, tell the Tech Pubs manager about your interest. Ask the manager if she will help you by mentoring you, training you, or even letting you do some work on your own time with the hope of eventually transitioning to that department.

> **TRUE STORIES**
>
> At two companies I've worked for, ambitious administrative assistants who wanted to be technical writers asked for help and mentoring. They volunteered to do Tech Pubs work on their own time, which was a fair trade for the training my department was able to provide. And then when there was room for an opening, guess who got hired?

If your company doesn't have a Tech Pubs department, maybe you can fill a tech-writing need your company doesn't know it has. Does your company hire freelance writers, or do developers and quality assurance people write user documentation? Could your company benefit from technical writing but doesn't want to spend extra money to hire anyone now? See if you can volunteer to do some of this work. Keep your eyes open for opportunities and you'll be surprised what you find.

Leverage Your Business Knowledge

Not all technical writing jobs involve software. There are technical writers in many fields, so no matter what your current business expertise is, it can be useful in your new career. Try to figure out a way to combine your business expertise with writing. If you worked in the medical field, or finance, or manufacturing—or almost any field—there are likely to be tech writing jobs in those fields.

When looking for a job, try to find a company that would be interested in someone with your particular business expertise, and then make sure you emphasize any writing experience gained on the job. Anything in either the writing or technical arena is going to help you in your new career as a tech writer. Written personnel guidelines? That's procedure writing. Helped your boss edit presentations? That may be technical editing. Developed flowcharts? That's related to document design and procedure analysis. Taught English? That can be instructional design and training.

Emphasize everything you've ever done on any job that relates to technical writing, from flowcharting to writing reports. As you continue to read this book, you'll find more to add to this list. Combined with the technical training and tools experience you'll be acquiring, you will increase your chances of being hired.

Scope Out the Possibilities

Look around you at the companies that hire or might hire technical writers. Find out what kinds of skills they seek in the programmers they hire. Are they looking for people with C++ and Java? Networking? SQL? Linux operating systems? That can give you an idea of what kinds of expertise will get your foot in the door.

And it gives you one other advantage—companies that hire software engineers also hire tech writers. Let them know you're available while you're building up the skills you need.

Create Your Own Intern Opportunities

It's never too late to become an intern. I know many career-changers who enrolled part-time in a certificate or nighttime degree program and became technical writer interns—I've even hired some of them. Some of those people found the intern opportunities first while they were job-hunting, and then went and enrolled in the programs to make themselves eligible to become an intern. Others were in an evening technical communications program and then sought out intern jobs.

People like this make very good interns, because they have years of business experience already as well as the drive to break into the field. You could try to do this yourself by finding an intern job opening and telling the hiring manager that you will enroll in a course to make you eligible for internship. If your previous job was in anything that the new company would find useful, combining that with technical writing coursework can be a great match for a company that is seeking an intern.

Why Is It So Hard to Get an Interview?

So let's say you've read this book and learned it by heart, you've learned some tools, taken some technical courses, and beefed up your resume. And yet you still can't get an interview. What gives?

Hiring managers are under enormous pressure to produce. They are doing much more with many fewer employees than they ever did before and the image in their minds of the employee they want looks more like a hired gun than a kid with a slingshot. And even though you might be the perfect job candidate, they don't know that...yet.

Before you start the job-hunt, learn all you can from veteran talent recruiter Nick Corcodilos at **asktheheadhunter.com**. On his website, blog, and in his books, Nick talks about why bombarding hundreds of companies with resumes does not work, and what you can do instead.

You'll never look at job-hunting in the same way again after you learn from Nick how to focus on what the job search is all about: showing how you can do a job profitably for the employer.

Many managers spend a lot of time looking for someone who fits their idealized wish list. Then as they start to review resumes, they discover that no one—or at least no one they can afford—fits every characteristic they are looking for. And sometimes they find that the applicants that look great on paper don't look so great when they're sitting across an interview table. It can take a long time and a lot of effort to find the right person. In fact, employers often spend months looking for the "perfect" candidate, when they could have spent those same months training a close-to-perfect candidate.

Network, Network, Network

As I said earlier, most jobs are filled because of personal connections and not from sending in a resume? Because of this, it's your responsibility to make sure you meet as many working tech writers and hiring managers as you can, as well as other people who work at the companies where you want to work. The more people you talk with, the better your chance of meeting the manager who will realize that you have the desired qualities of a tech writer.

How do you meet real live working tech writers and hiring managers? Read on.

Join STC

The Society for Technical Communication (STC) is the most important and largest organization for technical writers. STC (**stc.org**) is a professional association that advances the arts and sciences of technical communication. A new or would-be technical writer *must* join this organization.

STC has more than 150 chapters worldwide, and there is likely to be one near you, where you'll be able to attend meetings and hear presentations on important subjects. You'll be able to access members-only job postings. You can enroll in web seminars and courses, some of them free. You can also join one or more of its Special Interest Groups (SIGs) to connect with a virtual and real community of fellow tech writers interested in topics such as information design, international technical communication, usability, and more.

The most important thing to remember about networking is that you must start doing it *before* you need it. It can take time before the effort you put in pays off.

Your STC membership gives you access to an extensive jobs database and a worldwide community of other technical writers. You will learn from STC publications, websites, and seminars. You can enter your documentation in STC-sponsored competitions, where winning will help gain positive recognition at work. You can apply for the Certified Professional in Technical Communication (CPTC) credential.

Make the most of your time at chapter meetings. Be friendly and professional and meet as many people as you can. This is no time to be a wallflower. If you meet hiring managers, talk to them! Even if you don't happen to meet a hiring manager at the chapter meeting, you can meet the next best thing—working technical writers who might help you find your next job.

Scale the Summit

While chapter meetings are probably the best place for you to network, also good is STC's yearly conference, the TechComm Summit. The Summit takes place every year in the US and attracts technical communicators from around the world. You'll be able to network for several days (longer if you sign up for pre-conference sessions) and learn a lot, too.

The employment corner at the Summit is an excellent place to look for work. If you go to the conference, be sure to stop by there—but that's not where you're likely to strike employment gold. Keep in mind that the Summit attracts many, many newcomers, all eager to break into the field, and many of them are at the conference trying to find work. It's easy for one more to fade into the crowd.

Stand out from the crowd

Do what you can to meet the people who are in a position to hire. Do something (pleasant, please!) to help them remember you. If you spend the whole STC TechComm Summit attending "Lone Writer" or "How to Write Your Resume" sessions, don't be surprised when you leave the conference still alone, with a great-looking resume that nobody has seen. Do what other successful beginners have done to make yourself stand out:

- **Don't be shy.** Many technical writers are a bit introverted and it can be hard for them to strike up conversations with strangers. If this describes you, however, remember that being able to talk to a number of people is going to be part of your job, and now is as good a time as any to practice. Exchange business cards with people you meet and don't hesitate to ask if you can join a group for lunch. As uncomfortable as it may be, do it.

If you have difficulty speaking to people you don't know, practice your introduction several times in your hotel room the night before your first event. Then, when you're sitting at lunch at a table of strangers or next to someone at a session, turn to them before you lose your nerve, say "Hi," and introduce yourself. It gets easier with practice. It really does.

- **Be who you aim to be.** Think of yourself as a technical writer, not a "wannabe." Dress and act like a professional (this is a trade conference, not a vacation) and carry business cards with your name, email address, and your website. (Yes, you need a website.) Make sure the cards include the title "Technical Writer." Pay attention to the details of how you present yourself, including showing up on time to sessions and meals.
- **Attend the sessions aimed at managers.** Go where your next boss might be. Sessions for managers are clearly marked on the program. Don't just attend and sit quietly in a corner—talk to people while you're in these sessions and try to set up time to talk with them later to go over some of the ideas that came up during the session.
- **Make a lasting impression.** Make sure the people who count remember who you are. Collect cards or email addresses from the people you meet and connect with them later, both during and after the conference. Don't hound them for a job—instead, send them information they can use. Forward them articles of interest or follow-ups to conversations you had at the Summit. Get them thinking about you as a colleague and maybe you will become one.

INSIDERS KNOW

Before you start participating in the TECHWR-L mailing list community, read the messages from fellow members and read some of the archives on the website. You'll find that people on mailing lists are willing to offer help, but not if they feel you are trying to get them to do your work. Don't ask for advice that can be answered with a simple search of the archives.

And don't get into squabbles with anyone online. Remember that your postings will live on. Hiring managers often do a web search on candidates before they invite them in for an interview or before they make an offer.

TECHWHIRL.com and the TECHWR-L Mailing List

You can take advantage of the extensive technical writing resources at the TechWhirl website at **techwhirl.com**, the "Online Magazine & Discussions for Today's Tech Writer." Here you'll find features and articles, columns, access to a job board, and perhaps the most valuable resource for you, the TECHWR-L email discussion group.

Imagine a community of fellow technical writers the size of a small town, all ready to chip in with advice and commentary. The TECHWR-L email discussion group has been active since 1993 with over 3,000 members giving their opinions and sharing facts on the topics that matter most to technical communicators.

Newcomers to the field have found everything there from a warm welcome to a heated argument, once they dare to get out of "lurk" mode and jump into the discussions. You'll definitely want to join to see what the buzz is all about.

Join Online Networking Sites

It's important to take advantage of online resources for networking. LinkedIn (**linkedin.com**) is, at the time of this writing, one of the best places for you to do that. With more than 100 million members, LinkedIn can be a great source of contacts for you.

If you can afford it, upgrade to a business account. This gives you access to more people within the companies you want to work for.

But it's not as easy as putting your profile up and sitting back and waiting for recruiters to find you. You may have to look for some creative and more proactive ways to use LinkedIn to your advantage.

LinkedIn has many professional user groups. Join the ones that interest you and participate in the discussions that arise, so your name will become known to the group members. Just remember that your postings can live forever, so always act professional.

Use LinkedIn to find people you need to know. If you're interested in working for Acme Widgets Corporation but don't know anyone at the company, you can search on LinkedIn to find someone who works there. There's no harm in reaching out even to someone who is not in your immediate network, to ask about the company, to find a connection in a particular department, or to try to gain an informational interview.

Seek Insider Information

So what ***about*** the informational interview? Some experts think it's a waste of time; some think it is useful. I think it can be a useful networking tool, and have hired people I met during informational interviews. As a tech writer trying to break into the field, you need to meet as many people as you can in the companies you want to work for and in the field you want to work. People like to help, and many hiring managers are happy to take 15 minutes or half an hour out of their day to talk to you.

Keep your request professional and make sure you let them know there is no pressure on them to hire you. Keep the meeting short—no more than half an hour—and make the most of your time. Even though this is not a job interview, put your best foot forward as if it is.

The logical next step from informational interview is real interview, and if you follow the advice in this book, you have a very good chance of getting one.

> **INSIDERS KNOW**
>
> Impress potential employers with your tools knowledge: create an ebook out of one of your portfolio samples and display it on one or more devices. Learn more about ebooks and formats required in Chapter 12, "You Want it *How?*"

Winning Interview Tips

There's not much I can tell you about interviews that isn't covered in a thousand other books, but there are a few things you should know about technical writer interview etiquette for that time when you do win an interview with the company of your choice.

When you get an interview, prepare for it. This means learning as much as you can about the company and its business needs, the people you will be meeting, and the value you can bring to them.

Passing the Phone Screen

Many first interviews are done on the phone. Without passing the phone screen, you won't make it to the face-to-face interview phase.

It can be much more difficult to sell yourself by telephone, and applicants are often filtered out because of lackluster phone skills. Have your resume, a list of questions, and the company's website up on your computer before the phone call. Amp up your level of enthusiasm and animation, since you won't be able to use body language and your portfolio to let the interviewer know what a good candidate you are.

The Video Interview

Many companies, especially those with remote team members, interview with webcams or teleconferencing equipment. For you, this requires you to look as good as you do at an on-site interview (at least from the waist up!) and be as enthusiastic and energetic as you must be during a phone interview.

If you are asked to do a video interview, practice first by talking about yourself in front of your webcam. This will help you to see if you need to smile more, hold your head differently, or move yourself to an area with a better backdrop.

The On-Site Interview

Bring copies of your resume with you in case your interviewers don't have one. And definitely bring your portfolio. If your interviewers don't ask to see samples of your work, it's your responsibility to make sure you show them.

If your portfolio is online, send the link to the interviewers in advance, but still expect to show your work at the interview. Consider bringing printouts or hard-copy versions of some of your materials in addition to your laptop. Many interviewers like to leaf through documentation as they speak with you.

Not all companies allow Internet access to guests, so make sure all of your portfolio is on your local drive as well as online. Be ready—if you have a slow computer, turn it on while you are in the parking lot or waiting for the interviewer. There's nothing more embarrassing than fumbling around with your equipment while the hiring manager stands by looking at his watch.

> **INSIDERS KNOW**
>
> Thanks to the Internet, you can learn a lot about the companies you may work for. Obviously, the company's website is your best starting point. Here you'll find information about the company's products, management team, press releases and more.
>
> Before you go to an interview, ask whom you'll be speaking to and what their positions are. It doesn't hurt to do a search on them to find out what's important to them, although it *can* hurt if you are overly obvious about your search. You don't want to make your interviewers feel uncomfortable by asking personal questions or being too obvious about your search.

Bring a notebook with you that contains your questions about the company, the job itself, the department, and the people with whom you'll be working. You'll be asked repeatedly if you have any more questions, so it's a good idea to bring a ready-made supply with you. Writing them down beforehand means you won't forget them. Take notes, but not so many that you are too busy writing to speak, and definitely be careful that you aren't so busy writing notes that you don't hear what the interviewer is saying.

Be nice to the receptionist and the recruiters and anyone you meet along the way. It is unfortunate and surprising that job-seekers sometimes are less respectful to the people whom they feel are not important. Don't forget: those people often give input to the hiring manager.

> **TRUE STORIES**
>
> Deanna's success story demonstrates nearly all of the principles discussed in this chapter. She was laid off after 22 years in a different career. As she began job-hunting, she realized all of her roads had led to technical writing when she saw the common theme that had run through all of her professional and volunteer jobs—there was always a writing element that she had enjoyed and done well.
>
> While searching for a new job, Deanna continued to do volunteer writing work and took a six-month online certificate course in technical writing. She used her student discount to buy the most current versions of Adobe Creative Suite and FrameMaker. She reworked her resume, pushing "technical writer" more to the forefront each time.
>
> An ex-colleague forwarded her newly tailored resume and his personal recommendation when a job as technical writer opened up at his company. Deanna took a writing test (you have to love writing tests for a newbie or career-changer), aced it, and is now happily employed as a technical writer.

Presentation—It's More than the Portfolio

Remember to look professional while job-hunting. You may have the impression that everybody in high-tech wears shorts, T-shirts from companies that have gone out of business, and sandals. Not so. Or at least, not at the interview—and probably not on the first day of work. If you're unsure of how to dress, ask someone who works where you're interviewing—and then dress one level up. You don't want to scare a super-casual team by dressing like a Brooks Brothers mannequin, but you don't want to look like a slob, either. People notice both.

Seal the Deal

A special reminder to you soon-to-be tech writers: remember to try to close the deal. If you land a face-to-face interview at a company, don't leave without clearly telling the interviewers that you really want the job. Follow up with a written or emailed thank-you note in which you remind them how interested you are and how much value you can bring to the company. Include a link or reference to an article or conference topic that you think will be of interest to them—remember you want to be a colleague, not just a supplicant.

Little Things Mean a Lot

Does all this sound daunting and like a lot more work than you expected? It's true that breaking into a field can be difficult when you don't have related job experience. But the fact that many tech writers have come into the business through unexpected routes means that there isn't always a template for what their backgrounds should be.

The good news is that everyone I know who really wanted to get into technical writing has done so. Yes, some of them were lucky enough to break into the market during boom times, but others were not, and they used some combination of the methods described in this chapter.

Whom you know can really make the difference. If five people apply for a job and their qualifications are roughly equal, the one who is known to the interviewers has a better chance of getting the job, even if that relationship is through STC or a group like LinkedIn.

And *what* you do can make a difference. If you are the only one who sends a thank-you email afterwards and that email contains information useful for the hiring manager as well as a reminder that you want the job, you have a much stronger chance.

Checking up on the hiring manager's decision can help, too, but it requires delicate timing and the knowledge of how to stop just short of being annoying. Bear in mind that the way you show your persistence to the hiring manager demonstrates how you'll show it on the job. Persistence is an important characteristic that makes a good technical writer—but so is tact. Managers want employees who will doggedly go after information until they get it—while not making any enemies or causing trouble.

Everything counts in the job search. If you are the only one who follows up a week later with a polite request for status and a reminder that you are interested, this can also work in your favor.

It can take a while for you to find the person who will give you that first tech writing job. But it can be done. Don't give up, and do "work smart." Keep honing your skills and make sure you are in the public eye, on forums like TECHWR-L, LinkedIn, and your local STC chapter. Eventually, a hiring manager will realize that you've got what it takes to do the job and to do it right.

Part 2. Building the Foundation

You can't start writing until you know what you should be delivering. This section talks about what goes into making good documentation.

Chapter 5

How to Write Good (Documentation)

How to recognize good—or not so good—documentation when you see it, or when you write it yourself.

What's in this chapter

- The five keys to good documentation
- Ways to ensure accuracy
- Avoiding incompleteness
- How to recognize usability
- Some tips on clarity
- How to make sure your consistency is anything but foolish

To be a good technical writer, you have to produce good documentation. It sounds pretty straightforward. But how do you know if the documentation is good, mediocre, or a failure?

The quality of documentation is actually in the eye of the beholder. That is, different users have different needs, and if you haven't met the need of a specific user, your documentation is no good for that person. And if that person is representative of most of your customers, it's possible that the documentation is not good for anyone. I discuss that finicky user in Chapter 7, "It's All About Audience."

No matter what user you are writing for, there are certain attributes that all documentation should have before it can be considered "good." In roughly this order of importance, good documentation is all of the following:

- Correct
- Complete
- Usable
- Clear
- Consistent

There are other important characteristics of documentation, certainly. Design, discussed in Chapter 20, "Design and Layout," is important. Positive out-of-the-box experience, discussed in *Chapter 7, "It's All About Audience,"* is important. Context-sensitivity and searchability, discussed in Chapter 12, "You Want it *How?*" are important.

But those are all the icing on the documentation cake. Documentation doesn't need those things to be *good*. But it may need them to be *excellent*.

Correctness Is Key

What counts more than crisp writing, good grammar, accessibility, and beautiful design? Correct information, of course! If there is any single thing that can be said to be the most important characteristic of good documentation, it is correct—that is, error-free—content. Your user needs to trust in the document's information.

Nothing shatters trust faster than your reader discovering—always too late—that a crucial piece of information, confidently accepted, is wrong. The customer's business may depend on it. A life—if you are documenting medical equipment or heavy equipment or electrical work—may depend on it. Less seriously, but still important, the user is depending on documentation to be correct so she can perform a task, or entertain herself, or do her job.

The Cost of Incorrect Documentation

When a customer calls with a complaint, solving that problem takes time, and for a business, time equals money. It takes time for the customer support representative to take the call, chat, or email, and resolve the problem. The support rep uses still more time to write up the trouble ticket that makes its way back to you. You take valuable time away from other projects when you make the correction and issue a revision.

What hurts more than the time and expense is the bad reputation this can cause your business. People talk—in forums, on consumer sites, and during normal conversations—about what they don't like. And if documentation is misleading, wrong, or just plain bad, the word will get out there and reflect badly on your company.

It's surprising but true that you, a technical writer, can have such an effect on your company's bottom line.

How to Make it Right

There will be times when something goes wrong in the documentation and you feel the wrath. You might feel as if you don't have enough control when it comes to accuracy. After all, you didn't write the code or design the product, right? And you know the reviewers barely looked at your drafts when you sent them out.

But don't get defensive. Resolve the issue professionally and quickly, and try not to make any more mistakes. Here are some ways to make sure your documentation receives an A for accuracy:

> **TRUE STORIES**
>
> In 1998, the National Transportation Safety Board blamed General Electric for an engine fire on an American Airlines flight from Puerto Rico to Miami. The maintenance manual for the engine incorrectly described how to install one of the bolts. Not all documentation errors can cause such disasters, but they can cause big problems for business.

- **Know what you're writing about.** The best first line against incorrect information is for you to know yourself if what you write is correct. Chapter 10, "Become Your Own Subject Matter Expert," gives you some ideas on how to do this.
- **Keep thorough notes when you gather information and refer to them as you write.** Whether your information-gathering tools are digital recorders, laptops, mobile devices, or pen and paper, they all work as ways of recording information. It's not a bad idea to keep everything until the next release comes out. If you are ever in a position of having to defend yourself, it can't hurt to have all the facts. See Chapter 14, "Gathering Information," for more about this.
- **Have your drafts reviewed by people who know** whether what you've written is correct. Chapter 16, "Everybody's a Critic—Reviews and Reviewers," guides you in working with reviewers.

Completeness Counts: Make Sure Nothing's Missing

Incomplete documentation is bound to cause customer complaints. Incompleteness is just another form of inaccuracy when you omit a crucial step from a procedure, forget to tell users how to do something they need to do, or don't include an important topic.

Complete documentation includes everything the users need. If they need troubleshooting information or an explanation of error codes or a specific step in a procedure and this information is not there, then the documentation is incomplete to them. It doesn't matter to the customers that your team hasn't had the time or resources to do this job, or that your manager said it was a low priority, if it is mission-critical for them.

What's Not There Is Hard to See

It's not always easy to make your documentation complete. Your expert reviewers may confirm that the information you write is correct as it stands, but they don't always think about what's *not* there.

The origins of the term *stakeholder* refer to a neutral third party who holds onto money or property while its legal owner is being determined.

In today's business world, a stakeholder is a person who has a legitimate interest in the outcome of a project. Make sure you know who are the stakeholders in your projects.

This is one reason why experts don't always make the best technical writers. They are so familiar with the subject matter, they often don't notice what's missing. Their minds fill in the missing gaps. It's up to you, the technical writer, to make sure that all the information that needs to be documented *is* documented.

Up-front planning is the best way to deal with this type of completeness, with the aid of documentation plans and outlines as discussed in Chapter 9, "Process and Planning." When you and the stakeholders work out the documentation needs of your external and internal customers, you will also prioritize those deliverables according to their importance and the availability of your team's resources. As customers ask for additional documentation, you can add those requests to the plan and reprioritize as needed.

Cross-Reference to the Information They Need

Make sure you include obvious ways for the users to find the information they need. You don't get points for creating much-needed troubleshooting content or instructions on how a customer can return an online purchase if there's no way for the user to find it.

In manuals and online help, this means providing hyperlinks and cross-references wherever they are needed, to point the user to related information. It also means posting or distributing that information to a place where the user can obtain it.

Having a Handle on Usability

Usability is a word you've probably heard a lot. It means making products easier to use, by matching their behavior to user needs and expectations. There may be a Usability, or User Experience, department at your company, and you may even report into it.

Usability of a product typically boils down to being able to answer "yes" to four questions. Documentation can play a part in all of these areas.

- **Is it easy to use?** Wizards, contextual help, and easy-to-understand instructions can assist in ease of use.
- **Does it anticipate the user's needs?** Help, pop-ups, and on-screen text can guide the user in the right direction and toward a successful conclusion.
- **Can a user recover quickly from errors?** Step one in error recovery is to have an error message that is easily understood. Good error messages not only tell the user what the problem is, they also provide instructions on how to recover from the error state.
- **Does it provide feedback?** In other words, is the user notified when he either succeeds or fails? Is it clear when the process is completed?

Usability is the practice of taking human physical and psychological requirements into account when designing programs and computer content. This creates a better product and one that is more intuitive for the user. But it's not only an act of altruism—improving usability reduces business costs by cutting down on the number of calls to customer support and maintains and grows market share by creating loyal customers.

When a product is not usable, the tech writer is often expected to work miracles to explain it. We've all been there. And it's not pretty.

To assess a manual or help library's usability, make sure the documentation has been found to be acceptable on the following points:

- The user is able to find information quickly.
- Instructions are clear and easy to follow.
- Instructions work as described.
- Users can find their place quickly when coming back to it.

You may want to present information in various levels, allowing different levels of users to obtain more information as they need it. Mouseover or hover help adds more details when the cursor passes over various parts of a web page or a software screen. As a final step, you could provide a full page of detailed help content, anticipating questions a user has. Make sure the help is searchable so users with less frequently asked questions can still find their answers. As you can see, there's a lot to think about where document usability is concerned.

Chapter 7, "It's All About Audience," goes into detail about writing for the user.

Usability Is Also All About the Company

Usability isn't just about the user. It's also important to the company. The more usable a product, application, or service, the more loyal its customers, and the more profitable it is.

Because websites and e-commerce sites bring in the cash, many companies devote a great deal of money and time to making sure they are usable. Usability efforts can range from *heuristic evaluation* (evaluating a product against standard usability best practices) to monitoring and recording users as they perform tasks with the product or website.

Some companies spend much time doing *user-centered design*—design that is tested, revised, and tested again with surrogate customers until the majority of test subjects succeed at completing the task or tasks. If your company does this, make sure that your content is included and usability-tested as well. If your company does not do usability testing, ask someone who resembles your target customer

Morgan's story: I worked on several enterprise products, and our group worked closely with the User Experience team. I was invited to attend user research studies, and observed how users interacted with the products. This gave me insight into the information search strategies that real users used to get the information they needed. Boy, was this an eye opener! There's nothing like watching users struggle with the information you produce. It gave me empathy and forever changed the way I create documentation.

to follow your procedures exactly. You'll learn a lot as you watch.

If you are documenting a web-based application or consumer website, it's important to consider the ways in which your users will interact with the web pages. On an e-commerce site, for example, it is critical to prevent drop-off. You must keep the customers on the page as long as they need to be there and move them through the process until they complete it. Your page content should explain things to them clearly and succinctly. If they need further explanation, make sure the explanation is easy to find and does not break their flow. Everything within the pages must be designed to continue to move the customer forward.

> **TRUE STORIES**
>
> My User Experience team was once conducting usability tests for a redesign of a website at which users enrolled for a service. The procedure was complex and the service was complex, so the site had large amounts of text explaining what was going on and what the various options were.
>
> We discovered that the participants simply did not look at anything other than the first and last sentences in our carefully constructed paragraphs. By the end of the process, the majority had not learned what we wanted them to know.
>
> We resolved the issue by limiting all paragraphs within the flow to two sentences. To draw their eyes to an important point, we added bold text—emphasis—which definitely caught their attention.

Although you may be accustomed to writing long, detailed technical documentation, when writing web content brevity can be much more important. Consumers who are attempting to buy something online don't want to wade through long amounts of text to find out how to do it—they want to move quickly through the process. And your company doesn't want them to delay, because a customer who isn't moving ahead is one who is likely to drop off. If it is difficult to buy something on an e-commerce site, use a mobile device, or work with a software product, a consumer simply walks away. Worse, she might complain about the product on one of the many consumer websites, preventing many other potential customers from buying.

Clarity Is in Good Writing

Correct, complete, and usable information aren't enough if your writing is poor. Effective documentation must be well-written, and for technical writing, that doesn't mean flowery sentences or elegant variation. It means *clear* writing.

Good technical writing, unlike good fiction writing, presents content in a way that draws no attention to the language or the person who wrote it. You don't want your readers stopping to reflect upon your turn of phrase. This is a distraction when their purpose is to learn or act on the subject of your content.

Here are two key things to strive for to make your writing clear:

- **Eliminate unnecessary words.** "Click the **Add** button in order to create a new user account" can be written as "Click **Add** to create a new user." The fewer words, the easier your content will be to read and the less ambiguous your meaning will be.
- **Use short words and short sentences.** It is said that the average American adult reads at between an eighth-grade and ninth-grade reading level. (Depending on your target audience and the topic of your documentation, you may be able to go higher. Know your audience.)

Chapter 6, "Best Practices Make Perfect," goes into more detail about clear writing and reading levels.

INSIDERS KNOW

One thing that can confuse a tech writer who's trying to be consistent is what to call the products you're writing about. You'll often hear three or four names used for the same product.

A product in development often has an internal code name (a name used to prevent competitors from learning about the product while it's being developed) as well as the commercially used name. In addition to that, there might be a number of abbreviations used internally. Often the product name is changed at least once before it goes to market, yet the old names live on.

Before writing any customer-facing documentation, make certain you understand what the product is to be called.

Consistency—Is It Really the Hobgoblin of Little Minds?

In fact, what Ralph Waldo Emerson really said is, "A *foolish* consistency is the hobgoblin of little minds." By adhering to standards of consistency in your documentation, you'll be anything but foolish.

In conventional English, there is an abundance, a plethora—in fact, a myriad—of synonyms for many words. What this means is that the same, or nearly same, meaning can be expressed with a wide variety of different words.

Some writers think it's boring to use the same term over and over. Maybe in fiction, but this type of thinking doesn't apply to technical writing.

Sameness Is Not Dullness

Good documentation always uses the same word or term to mean the same thing. Every time.

You don't gain points for thinking of different ways to refer to an item. Sometimes a tech writer isn't deliberate when he uses different terms for the same thing. This can happen when a tech writer hears something called an internal name or nickname, or when the tech writer inserts material written by someone else, or when the writer acquires information from more than one subject matter expert. Unsure of the difference himself, the writer tries to cover all possibilities by leaving all the names in place.

Bad idea! If you don't understand it, it's a sure thing your user won't understand it. Using six different names to mean the same thing can only create confusion. Each time a new term is introduced, the user will think (sometimes rightly, sometimes wrongly) that it refers to something different.

The rules of product names are rarely determined by the engineering team, many of whom will continue to use internal names for the life of the product and in the source code. Product Marketing, Product Management, and Legal usually determine not only the official names of products but the rules for how you use them.

Good documentation is not only consistent in its terminology, it is also consistent in design, language, iconography, and typographical conventions. And this happens not only within a single document or help project or website, but also across all of your company's documentation.

If your release notes contain prerequisite information for product A, you don't want product B's prerequisite information in the installation guide, and product C not to mention prerequisite information anywhere.

The user should not have to learn things twice. If bold monospaced font is the typographical convention for user input in one document, then bold monospaced font should mean user input in every document. And in help. And on your web pages. If variables are indicated with italics in half your documentation, don't decide to use angle brackets in the other half of the documentation. Consistency...

Style Guides: Not Just a Fashion Statement

You want customers to buy more than one product from your company, don't you? All of your documentation should look as if it came from the same company, with the same imagery, branding, and terminology.

The best way to ensure consistency is to have a style guide—and then to follow it. A good style guide should lay down rules for everything from the meanings of specific terms, the use of typographical conventions, accepted spellings, and just about anything you can think of. See Chapter 18, "The Always-in-Style

Guide," for detailed information about style guides, including how to create your own.

Follow the Bouncing Word

By now, you should understand why consistency is so important. Forcing the reader to play guessing games is never a good idea. There's no need to make the user play a round of "What's My Definition?" as you try out one word and then another to refer to an object or action.

The QuickandDirtyTips.com site is dedicated to "Helping you do things better." One of its most useful and fun sections is Grammar Girl, Quick and Dirty Tips for Better Writing. The site contains tips on everything from word usage to spelling to punctuation. Check it out at **grammar.quickanddirtytips.com**.

What if you had to use or install a complex and expensive product all on your own, and the documentation used five different names for what appeared to be the same thing? That's the situation countless technicians and administrators find themselves in every day when using documentation developed by writers who don't understand the terms in the documentation they write.

Consistency allows the users to form patterns in their minds and save time because they don't have to think anew every time they see the pattern. Set up the expectation, and then meet that expectation. That's not boring, it's reliable.

Chapter 6

Best Practices Make Perfect

Solid practices that anyone can follow to write good documentation.

What's in this chapter

- The secrets of successful technical writers
- The informal style of today's technical writing
- A list of best practices that should become second nature

You may have wondered how it is that tech writers just seem to *know* things like, "write in the active voice," "use bullet lists," or "don't use the future tense." It seems that these *best practices*—methods or techniques that have consistently shown superior results—have become part of the tech writer's DNA.

Before the advent of the personal computer, technical writers wrote in a formal style for other technical experts who understood the language. When computers began to be commonly used in the workforce, and even more so when they became something an average person could buy for their home, the technical writing field changed in a big way.

Writers who had the gift of language rather than technical expertise were recruited into high-tech companies with the intention of creating more user-friendly, easy-to-understand documentation that would help new users understand computers and software. Major corporations hired *human factors* (usability/ergonomics experts) evaluators to provide recommendations on everything from type size to the number of steps in a procedure.

The technical writers who learned in those companies moved on to other companies, trained beginners, or started schools or consulting companies or recruiting companies themselves. They trained the technical writers who got their jobs during the dot-com boom, and in a very short time, these best practices became part of the tech writing culture.

In this chapter, I'll talk about the best practices that you ought to know about and ought to practice as a technical writer. The knowledge of these guidelines can make the difference between exposing yourself as a novice and showing yourself as an experienced practitioner.

It's not only about what makes *you* look good—applying these principles will make your documentation better, too.

Best Practices all Point to Clear Writing

All of the standards in this chapter are based on an intention to make the content clearer for the user. Clear writing—writing that avoids guesswork and explains without ambiguity—is where the rubber meets the road. But being able to write clearly doesn't take long years of monkish self-discipline or top-secret neuro-programming. It's a learned skill that's based on some simple (and clear!) guidelines and some down-to-earth common sense.

The best practices of technical writing, useful as they are, shouldn't be followed blindly. I'll talk about some of the rules that have been questioned over the years as well as the ones that seem to have no right answer—principles that technical writers have been arguing about for years, each side convinced that it is right—and let you decide.

The bottom line is always going to be how well your user understands your content. If following a rule (including one you find in this book) means you have to write an awkward or unclear sentence, don't do it. If following a rule means that you have to spend a huge amount of time figuring out a way to follow that rule, don't do it. The return on investment just isn't there.

Understand your users, understand the information they need, and then modify the rules to work for your organization. And once you have them defined, add them to your style guide so they will be easy to find, easy to follow, and become part of *your* DNA. Style guides are discussed in Chapter 18, "The Always-in-Style Guide."

The Best of the Best Practices

The practices in this chapter should become part of you, as unconscious as your decision to use correct grammar when you speak, and as deliberate as your choice of the right tool would be for a home-improvement job.

And so, in something close to alphabetical order, here are the best practices every technical writer should know.

Active Voice: Don't Worry, Be Active

Active voice is one of the cornerstones of clear writing. Active voice means that the subject of the sentence (I'm sure you remember learning in grade school that the subject is what the sentence is about and the verb is the action word) performs the action, as in the following example:

A network client sends requests to a server.

The sentence is about a network client, so that is the subject. What's the subject doing? It's sending requests.

Unsurprisingly, the opposite of active voice is passive voice, in which the subject is the recipient of the verb's action.

Requests are sent to a server by a network client.

Now the subject of the sentence is "requests," and they not performing an action, but instead are being acted *upon*.

You can recognize a passive-voice sentence by the fact that the verb takes on some form of the verb "to be" (am, are, is, was, were) coupled with the past participle form. (If you still remember that from school, good for you!) In the previous example, the verb is *are sent* versus the simpler *send*. Software is installed. The system was shut down. Networks are configured. Files were copied.

Voice is a grammatical term that describes how the subject and verb in a sentence relate to each other. *Active voice* means the subject is doing the action of the verb.

Passive voice means the subject is receiving the action of the verb, with no indication of who or what is doing the action. Stick with active voice as much as you can so readers can clearly understand what is happening

Technical writers generally enforce the active-voice rule because the active voice is less ambiguous than the passive voice. It is very obvious who—or what—makes something happen in a sentence using the active voice, with little room for misinterpretation. It also uses fewer words, which enhances clarity.

Writing in the active voice can force a technical writer to understand the subject matter. Before you can explain something to another, you have to understand it yourself, and if you don't know what does what to what, you're not going to be able to write a clear sentence in the active voice.

It's not uncommon for a technical writer who doesn't really understand his subject matter to try to hide it by using the passive voice. If you find yourself fudging by using the passive voice because you don't know what is causing the action to occur, it's a cue you may need to learn more about your topic.

Passive voice, however, has its place. Use the passive voice when the subject is:

- Unknown (although the subject should be truly unknown, not simply because you don't happen to know it)
- Unimportant
- Less important than the receiver of the action

For example, "The content is backed up continually" is an example of a passive voice sentence that makes sense. The *content* is the important focus here. It wouldn't even make sense to say that a combination of software programs, operations people, scheduling commands, and machines all work together to back up the content.

Calling a Spade a Spade (Not a "Manually Operated Multipurpose Soil Manipulation Instrument")

When Mark Twain was writing for newspapers and being paid seven cents a word, he said this about using short, simple words: "I never write *metropolis* for seven cents because I can get the same price for *city*." Apply that same principle to your own writing—look for the short word gets the idea across, instead of a long one.

You don't get paid extra to write "utilize" instead of "use" and "vehicle" instead of "car." Technical writing isn't about showing off your vocabulary or looking in your thesaurus. It's about communicating a technical subject to a reader. Shorter, simpler words are easier to read and easier to understand. And that's important to know when clear writing is the goal.

However, you must also use the terms your audience uses, even if those terms are longer and more complex. Technical writers who don't fully understand their subjects sometimes make the mistake of trying to rewrite a technical term they aren't familiar with, rather than using the one currently understood by people who use the product. Use the terms your customers understand.

Dodging Bullets: Creating Effective Bullet Lists

Technical documentation includes many bullet lists, and as a new technical writer, you may wonder why. This section talks about some of the reasons to use bullet lists (presented in a bullet list!):

- They emphasize important information in an easy-to-read format.
- They create more white space so the page doesn't look like a solid block of text.
- They group together items of equal importance.

You should always introduce your bullet list. Some style guides require a full sentence as an introduction, the way I did it here. Others use a heading style. Still others are fine with a sentence fragment that ends with a colon.

> **INSIDERS KNOW**
>
> Eye-tracking tools that measure eye positions and movement are used often in web usability research and testing, and writers can benefit from the knowledge gained in these studies.
>
> Eye-tracking studies show that users' eyes are drawn to bullet lists and bold text. Effective use of these can make a big difference when you want to catch a reader's attention.

Bullet lists must always follow parallel form. That means that every item in a list begins with the same part of speech and has a similar format. If the first item in a bullet list uses a *gerund* ("-ing" word), every item should use gerund phrasing. If the first item is a question, they should all be questions. If the first item in a list starts with an *infinitive* ("to" verb), they should all start with an infinitive.

If you have only a single item, do not put a bullet in front of it. Use your standard body text format.

Bullet lists should be neither short nor long. Fewer than three items does not make a list, and more than six starts to become difficult to read and can cause the reader to lose her place. Do not overuse bullet lists, either. Too many bullets are definitely too much of a good thing, and they lose their emphasis.

Because items in a bullet list should always be of roughly equal importance, do not use a bullet list when you want to present a series of items that are meant to be listed in a sequence or order of importance or hierarchy. In a case like that, use a numbered list.

Do you know the difference between a note, a warning, and a caution?

A *note* is a piece of supplemental information that is not crucial, though it may be interesting. A user can skip a note without losing anything essential. A *caution* means there is possible damage to equipment. A *warning* means there is possible danger to a person.When applying emphasis to warnings and cautions, make sure you use fonts or colors that highlight the more important ones.

A caution should be emphasized with italics, bold, or larger type to indicate its importance. A warning should have the most emphasis, whether it's larger type, boldface, or an icon to draw the user's eye.

Make sure you follow your department's guidelines for how to punctuate, capitalize, and structure bullet lists. There are many ways to do it, and it seems every writer has a preference. I talk about some of the options in Chapter 18, "The Always-in-Style Guide." It's not so important which method you choose. But whichever it is, be consistent with this style.

Emphasis: I *Really* Mean It

There will be times when a word or phrase deserves special emphasis to attract the reader's attention. There are many options available for highlighting and emphasizing words and phrases with italics, color, underlines, or bold type.

And it's not limited to the typographical changes. You can use indents, *outdents* (a line that extends outside of the normal margin), rules, and even create a style complete with an icon, as in the following examples:

Do *not* touch a person who has come into contact with a live electrical wire.

Do <u>not</u> touch a person who has come into contact with a live electrical wire.

Do *not* touch a person who has come into contact with a live electrical wire.

With so many options comes the temptation to overuse them. Don't do it. It's much better to be sparing in your use of emphasis. Why?

Emphasis adds a visual element that the reader must interpret. This is a good thing when you want the reader to slow down and notice your emphasis. But if you overuse any type of emphasis, the reader starts to ignore it.

Emphasis is like speaking loudly. People jump when they hear a loud word the first time. But who listens to someone who talks too loudly all the time? After a while, people tune out the loud talker. And that's hardly what emphasis is meant to achieve.

A Foreign Affair

When writing in English, refrain from using obscure and unnecessary foreign words. That includes all Latin abbreviations such as *etc., e.g.,* and others. The table below shows words that you can use instead of some of the Latin abbreviations you may be tempted to write.

Instead of:	Write this:
et al.	and others
etc.	and others, and the rest
e.g.	for example
i.e.	that is

Gender-Neutral Language: He Said, She Said

The English language has the peculiarity—or convenience—of having gender-neutral plural pronouns (*them* and *they*) but no gender-neutral singular pronoun for individuals. For that, we are stuck with *he* and *she*. (Not a lot of people are asking to be called *it* at the moment.)

Don't think about applying the historic convention of using the male pronoun to mean a person of either sex. Although some people still argue in favor of doing so, times change, and so does convention, and convention today says not to do it.

But what to do instead? Writing "he or she" or "he/she" or "s/he" may be more politically correct, but it makes for some very awkward writing.

The Finnish language has only gender-neutral pronouns. The word *hän* means both *she* and *he*, and it's sometimes necessary when speaking to explain that you are referring to a woman or a man. The only differentiation of pronouns is between humans and animals—the word *ne* means *it* and refers to non-human subjects. However, language evolves, and casual speakers frequently refer to humans as "it" and pet-lovers might call their pets by the human pronoun!

Many writers avoid this problem by using the plural pronoun in all cases, even when it does not match the subject. They write something like the following mismatched sentence:

> Assign owner privileges to a user to allow them to modify their own data.

It's a solution, and one that is common today, but still awkward. Grammatical purists hate it too, and you want to avoid doing anything that calls negative attention to the language.

Don't worry—there are a few tactics that technical writers can employ to avoid the gender-neutrality problem.

In fact, it's possible that you will never have an opportunity to write in the *third person* (that is, designating a person other than yourself or the one you are speaking to, with pronouns like *he, she, they,* or *them*) during your entire tech writing career. Technical writing today typically speaks directly to the user (called writing in the *second person* with pronouns like *you*), so you will probably be able to avoid the he/she problem.

That's what I do throughout this book; I speak directly to you, the reader, and it's only when I refer to examples of users or other technical writers that I need to use a pronoun like *he* or *she*. When I have to use a personal pronoun, I more or less alternate the use of *he* and *she*.

One of the most effective solutions to avoid difficulties is to simply "write around" the issue. This means constructing your sentences in ways that don't need to include "him" or "her." It isn't as hard as you might think.

> Users with owner privileges can modify their own data.
>
> Assign owner privileges to users so they can modify their own data.

The most challenging part of nonsexist writing is simply raising your own awareness about it. Once you have that consciousness, dealing with it in your writing becomes easier and more natural.

Humor: Funny Is as Funny Does

Humor in technical writing is a funny thing. No, wait, that's not what I mean...

And that's the problem with trying to use humor—it seldom comes across the way you mean it.

Humor is highly subjective, and varies widely from person to person and from culture to culture. As I'm sure you know. Think of jokes that you find offensive or tasteless or just plain unfunny, yet there are people laughing loudly at them right now. There's just no accounting for taste.

Humor is rather boring, too, the second or tenth or hundredth time you have to read it or listen to it, and technical documentation is often meant to be used and referred to again and again. You don't want to do anything that could make your user grow tired of or irritated with the documentation.

The bottom line is that humor doesn't do a thing to improve clear, solid writing, which, after all, is your primary objective. So although the desire to entertain your users is an admirable one, I have to warn you that it is generally a mistake. Take my advice—please.

Imperatives: That's an Order

Technical writers use the *imperative mood* to write instructions. The imperative mood is the sentence form used to express commands, using the infinitive without "to." You speak directly to your users when you use imperatives, telling them what to do in very direct terms, with no excessive language. The imperative form implies the word "you" without including it.

For example:

1. Fill in the form.
2. Click **Next**.
3. Enter your password.
4. Click **Finish**.

That's a lot shorter than "The user should click the **Next** button" or even "You click **Next**."

Humor about technical writing is another thing altogether, and we could use more of it. Do a web search for *A Grandchild's Guide to Using Grandpa's Computer*, by Gene Ziegler (which is widely circulated as *If Dr. Seuss Were a Technical Writer*).

Gryphon Mountain Journals contains much information about technical writing, including cartoons, at **gryphonmountain.net**.

The early Franklin Ace 100 computer manual contained a lot of humor, including a chapter called, "The Ancestral Territorial Imperatives of the Trumpeter Swan." Do a search for that phrase—there are many copies of that manual online. It's a nice slice of history, but not necessarily one I'd urge you to try to repeat!

And you can always find Tina the Technical Writer at **dilbert.com**.

Writing instructions, or procedures, in this way assumes that the user will read these instructions and follow them in real time. Direct, unambiguous, clear, the imperative strikes the right mood.

KISS: Keep It Simple, Smartie

With technical subjects that are often complex and sometimes need long documents to explain them, it may sound contradictory to say "Keep it simple." But applying the KISS mantra as you write technical documentation will be good for your users. This is similar to what I talked about in "Calling a Spade a Spade (Not a "Manually Operated Multipurpose Soil Manipulation Instrument")" on page 58, but keeping it simple means reducing the number of words altogether as well as making your words short.

Keep it simple by following these rules:

- Use a short word rather than a long one where you can.
- Keep paragraphs to no more than six lines.
- Keep sentences to 25 words or less.

Unusual words and overlong sentences are the first causes of reading difficulty. You don't want to overload your reader with too much to process. In technical writing, small paragraphs—even single-sentence paragraphs if necessary—can make it easier for the user to absorb the information. You are not dumbing-down or condescending to your reader by applying this practice.

A good rule of thumb is to keep paragraphs to a maximum of six lines if possible. Not six sentences—six lines. If you can't get the paragraph's idea across in six or fewer lines, that's a clue that you need to look more closely at what you're trying to say. You may have too many words, or you may be trying to include two ideas in one paragraph.

Another advantage to short paragraphs is that it adds white space to your documentation. I discuss white space in Chapter 20, "Design and Layout."

Shortening for Simplicity

Keep an eye out for wordiness in general. As you edit and revise, you'll find that there are many situations where you can eliminate words without hurting the meaning.Take this example from a real interoffice memo:

> Should personnel in your department require additional assistance, please register a request with a Technical Support representative.

What it means, of course, is:

> For more help, contact Technical Support.

The rewritten sentence has eliminated 11 words and an amazing 27 syllables, greatly improving readability.

Writers know it is easier to write long, wordy content than to produce tight, succinct prose. The following quote has been attributed to Samuel Johnson, Mark Twain, Blaise Pascal, and probably many others:

> *I did not have time to write you a short letter, so I wrote you a long one instead.*

Whoever said it was correct, though. You may not always have time to tighten up your content when you're trying to get a release out the door.

The greater the number of words and the more words in a sentence or paragraph, the higher the reading level. And since a good tech writer strives for an appropriate reading level and clarity, get rid of those unneeded words and choose shorter, simpler words instead of long ones.

You'll find that there are many words and phrases that you can simply delete without losing meaning. At their worst, they restate the obvious; at their best, they add unnecessary clutter. Here are some words and phrases that can nearly always be cut out or shortened:

- **in order to.** Can be shortened to "to."
- **a large number of.** Can be replaced with "many."
- **in the process of.** Can be eliminated altogether.
- **there is/there are.** Can be eliminated.
- **in the event that.** Can be changed to "if."
- **definitely.** Can be eliminated.
- **actually.** Can be eliminated.
- **totally, particularly, especially.** Eliminate them!

The two publications with the largest circulations, *TV Guide* and *Readers Digest* are written at the ninth-grade level, which is what the average adult reads at. Tests show that people tend to read at two grades below their actual reading level when reading for recreation. This explains why the most popular novels are written at the seventh-grade level.

Other extra words appear as *redundant pairs*—words that people tend to put together out of habit, but actually one word is unneeded. "Future plans" can be written as "plans." "Final outcome" is simply "outcome."

I'm sure you can think of many more examples. Look for words and phrases that can be removed without changing the meaning of the sentence, and you'll be surprised at how many you'll find.

If you want to check the readability of your work, there are many readability formulas available to you. The Flesch-Kincaid reading index, which factors in the average number of words per sentence and the average number of syllables per word (along with use of the passive voice), is built into Microsoft Word. Other reading index tools can be found at **readabilityformulas.com**.

Learning from Simplified Technical English (STE)

Simplified Technical English (**asd-ste100.org**) is a controlled language used in aerospace maintenance documentation. Since aerospace maintenance mistakes can cost lives, STE strives to reduce ambiguity, improve the clarity of procedural writing, and make translation—both human and machine—easier and cheaper.

STE standards are like technical writing best practices on steroids, and they are enforced. The number of permitted words is limited, and words are given very specific meanings. For example, "test" is permitted as a noun but not a verb. "Burn," "burns," and "burned" are permitted, but "burning" is not.

The US government supports a standard called Plain Language which is not as strict as STE. Go to **plainlanguage.gov** to learn more.

You probably don't want to adhere to such strict standards as these, but you can learn a lot from them about consistent terminology and clear writing.

Negatives: Let's be Positive, People

Avoid using negatives in technical writing.

There are two reasons for this. First, using negatives can lead to misunderstandings and slow down a user's comprehension. When you tell the users what *not* to do, you cause them to pause for a moment while they decide what they are being told. You risk that they will retain the wrong information.

One negative often leads to another before the writer knows what's happening, leading to double negatives and unclear writing. It can be difficult to understanding the meaning of a sentence with a double negative, as seen in the following example:

The **Close** button is not enabled if you have not selected the correct item.

What? Two negatives in one sentence and confusion about what the writer meant to convey.

Another reason to avoid negatives is that they suggest something's wrong. Manufacturers like to have their products always referred to in glowing terms, so avoid those negatives!

As do most rules, this one has an exception. In cases of warnings or cautions in which you must let the users know that they should *not* do something, a negative is acceptable and preferable. Let's look again at the example given in "Emphasis: I Really Mean It" on page 60, where the emphasis of the word "not" is an important part of the statement:

Do *not* touch a person who has come into contact with a live electrical wire.

Present Tense: Be Here Now

For some reason, the word "will" makes its way into a lot of technical documentation. It's an easy word to delete if you write in the present tense. Assume your reader is following each step of the procedure while simultaneously performing the task.

You only need to write in the future tense when you are actually writing about something that *will* happen in the future; for example:

> This backup will take place after you complete the configuration and submit changes.

In general, though, omit the word "will" from your sentences. No need to say "The firmware will enter an idle state" when you can simply say "The firmware is in an idle state" or "The firmware enters an idle state." If the user clicks a button to get an expected result, just say, "The page appears," not "The page will appear."

Remember, your reader should be following along in real time—not wondering when a future event will occur.

> **COFFEE BREAK**
>
> George Orwell's 1946 essay, *Politics and the English Language*, contains six rules that sound an awful lot as if they were used as the basis for today's best practices for technical writers:
>
> 1. Never use a metaphor, simile, or other figure of speech which you are used to seeing in print.
> 2. Never use a long word where a short one will do.
> 3. If it is possible to cut a word out, always cut it out.
> 4. Never use the passive where you can use the active.
> 5. Never use a foreign phrase, a scientific word, or a jargon word if you can think of an everyday English equivalent.
>
> And the final important rule:
>
> 6. Break any of these rules sooner than say anything outright barbarous.

Second Person: Yes, You...

Address your reader directly by using "you." This is called addressing someone in the second person. This book uses the second person to address you, the reader (yes, you!), and that's what you'll do in most of the technical documentation you'll create. The second person is related to the imperative discussed earlier, so some of the points overlap.

Speaking directly to the reader helps clarify in several ways:

- **The document is shorter.** Writing in the second person uses many fewer words than writing in the third person.
- **You avoid the passive voice.** It's easy to avoid the passive voice when speaking directly to the reader.
- **Instructions are easier to write**. You can state the steps in a process directly and without worrying about the subject of the sentence.

- **The language is clearer.** The users don't have to stop and wonder who is supposed to perform the action.
- **You avoid dealing with the issue of gender-neutrality.** Speaking directly to the reader means you don't write in the third person and therefore don't have to worry about whether the subject of your sentence is a "he" or a "she."
- **You address the reader in a friendly way.** Your writing becomes more conversational and easy to follow, like one person speaking to another.

Is this informal tone really acceptable for technical writing? You bet. The detached formal style that still shows up in much scientific and academic writing is not used much in technical documentation in today's high-tech world. For one thing, it takes too long to read and understand. (Who has time?) For another, the tone of the industry as a whole is more business-casual than suit-and-tie.

Don't step in the "you-do," however. This sounds unpleasant, and it certainly is for the user. By "you-do," I'm referring to the excessive use of the word *you*.

"You delete your unused files." "You enter the username and password into a configuration page." "You click **Submit**." And so on. You can eliminate three words—all spelled Y-O-U—from these lines without changing meaning. And you don't need to clean your shoes.

The Serial Comma: Not So Dangerous

There is more than one way to use commas, but the serial comma is a preferred style because it eliminates ambiguity, and as you may realize by now, technical writers strive to eliminate ambiguity.

Of course, to eliminate that ambiguity, you have to know what a serial comma is. A serial comma is a comma that precedes the final conjunction in a list. Luckily, tech writers also have at their disposal the handy bullet list, as discussed on page 59. Bullet lists take all the worry out of having to wonder whether your serial comma is in or out.

Omitting the final comma in a series is risky because it can lead to ambiguity for the reader. Consider the following sentence:

> Behind these doors are money, a lady and a tiger.

Does this mean the lady and the tiger are together behind one door? (Let's hope not!) How many doors are there—two? Or three? Don't leave your reader in doubt, even for a nanosecond. Always keep that serial comma where you—and your reader—need it.

Should You or Shouldn't You?

You shouldn't. The word *should* is very difficult for a user to understand. The user isn't sure what she is being told to do—does *should* mean the statement is a recommendation? A good idea? Something that might happen if all goes well? Or is it a mandatory action?

Some writers don't like to use the word *must* (they think it sounds bossy). I have no objection. The word *must* clearly conveys that an instruction is mandatory. *Should*, by nature, is ambiguous.

When you find yourself wanting to use the word *should*, stop to consider what you really mean. Is this a required action, or something that is recommended? Does it refer to something that's an expected result? ("The Finish page should appear...")

If it's recommended but not required, tell the user so and explain why performing this action is beneficial. If it is required, make that clear. If it refers to an action that you expect to occur, say so. ("The Finish page appears...")

If you're not sure *what* the right word is, stop and find out before you pick a word. It could mean the difference between a happy user and one whose system just crashed!

Easy as 1, 2, 3: Writing Procedures

At some point—probably early—in your tech-writing career, you'll be expected to write *procedures*, numbered steps that guide a user through a process. If you've followed a recipe or assembled furniture or bought something online, you've worked with procedures.

Writing procedures is a lot harder than following them! All the practices described in this book will come together as you create good procedures. When you finish writing step 1, you have to immediately follow it with step 2. Did you use excess words? Cut them out. Any unclear terms? Change them. Any possibility of misinterpretation? Rewrite the steps until they are crystal-clear.

Every process has two points: a trigger and an end point. The *trigger* is what starts the ball rolling. It could be the last step in the previous procedure or it could be an introductory paragraph preceding the procedure.

The *end point* is the state or condition that happens when the procedure reaches its natural conclusion. And it should be a natural conclusion. The user should feel that he has succeeded at reaching a goal, even if it's a midpoint or milestone within a larger set of procedures. This might be a congratulations or a thank-you if the user is going through a procedure online, or a description or picture of the expected result if the user is going through a procedure by following written documentation.

Procedural Guidelines

The guidelines for writing procedures are not much different than the guidelines for writing other components within technical documentation.

- **Write in the imperative mood, or by issuing commands.** If you need to elaborate on a step, add more information below the command, or add a note below or to the side of the step.
- **Make sure each step actually involves something the user does.** "A dialog appears" is not a step; it's the result of a step.

  For example:

  1. Click **Submit**. A dialog appears.

  Or:

  1. Click **Submit**.

     A dialog appears.

  Not:

  1. Click **Submit**.
  2. A dialog appears.

- **Address only one significant action, or one short group of related actions, in each step.** It's a common mistake to try to cram too much information into a single step.
- **Make sure each step is an action large enough to be meaningful**. If an action is microscopically small, go ahead and combine it with a follow-up step. You don't want a procedure to be full of tiny fragments too small to be significant.
- **Provide obvious transitions or connections.** The reader must never wonder if a step has been left out. Repeat "landmarks" such as screen shots every now and then to keep the reader oriented.

Working with Long Procedures

One method I use that seems to work for very long sets of procedures is to have a numbered "step heading" that can be used to group substeps. The heading is bold and easy to find if you are scanning many pages, and the number of headings and substeps can be very high.

It might look something like this:

Step 1. Complete the first part of the process

Some descriptive text about the first part of the process goes here.

1. Do one thing.
2. Do another thing.
3. Do yet another thing.

Step 2. Complete the second part of the process

Some descriptive text about the second part of the process goes here.

1. Do one thing.
2. Do another thing.
3. Do yet another thing.

The Rule of Seven—Not a Best Practice

There's one best practice that can be discarded by today's technical writers. Technical writers have long held to the principle that procedures should contain around seven steps. This principle is based on psychological research published in 1956 that suggested people can remember up to seven items plus or minus two.

Because of the Rule of Seven, many technical writers take care to limit the number of steps in a procedure. If a procedure threatens to be more than nine steps (the "plus two"), the writer breaks the procedure into substeps or a number of short procedures. But is this concern even necessary?

In fact, users are not expected to memorize procedural steps; you'll typically expect them to read the steps, once, and follow along as they read them.

> **INSIDERS KNOW**
>
> User Experience guru Jared Spool writes and speaks about the "scent of information," the cues that get the user on the right track to finding the content they need. He says that it doesn't matter how long a process is as long as the user continues to move forward in a purposeful manner. The same can be said for technical procedures. Visit Jared's website at **uie.com**.

More recent research suggests that the actual number of objects a human can hold in working memory is more like three or four. You don't want to limit your procedures to three or four steps—you'd have to cut them off before anything interesting started happening.

Should *you* follow the Rule of Seven? It all depends on the subject and context, of course, and what works for the user. There's no harm in limiting procedures to seven steps, or even fewer, but in the end, you can make procedures as long as they need to be, as long as they are well-organized, easy to follow, and enable the user to move smoothly from one step to the next.

Just make sure the users can always find their place in the process. Long blocks of text, whether sections of them are numbered or not, can cause users to get lost and be unable to find their way back. If a procedure has a large number of steps spanning many pages, it's difficult for the person following the steps to remember where he left off if his eyes leave the page, unless there is something to help him find his way. Headers, clear beginnings and ends, visual cues, and different type faces can all help to mark the content.

Show Some Respect

At the core of most technical writing best practices is a single, essential kernel: respect for the user. If you respect the user, you don't feel the need to waste her time, tap-dance around facts, or fancy up phrases with the latest esoteric word of the day from your desktop calendar.

Speak to users as if they were sitting next to you, asking for help. Remember, that user is someone like you.

Chapter 7

It's All About Audience

Analyze the people who use your documentation so you understand what they need—and don't need—to know

---

What's in this chapter

- Discovering and defining your users
- Writing for users who know more than you do
- The types of documentation different users want
- The best place to get together with your users
- How to fake it if you can't meet the user

Before you can plan the content of your project, there's a secret that every good tech writer needs to know: *know your audience*. Once you understand the people you are writing for and what their needs are, everything will come easily. (Well, okay, that may be too much to expect—but it *will* be easier.)

Journalism has the five Ws, the questions you must answer to write a good story: "Who, What, When, Where, and Why." Technical writing has one important W: the all-important starting question: "Who?"

This chapter talks about ways to learn who the users of your documentation are and the types of documentation they are likely to want.

Who Will Read the Documentation?

Before you start working on the project, you first must find out who the intended user is. Ideally, your boss or the product manager has already given this some thought and can give you some ideas. Getting an answer to that question simplifies your job.

The answer, however, might reveal that there are many different types of users, each of whom has different needs for the product. You may have believed that the most important user is the *end user* (the person who actually uses the product), but there are often many more than that—all the people along the way who learn about, buy, install, and are responsible for getting the product to the end user. On top of that, each of these people could be brand-new beginners, or they could be experienced, much more experienced than you.

Writing Up

There *will* be users who know much more than you do about the business your product is designed for. Or they will be much more technical than you. (It can be intimidating, creating developer documentation when you aren't a developer.) Despite those differences, you can still create useful and meaningful documentation for these users.

eBay is a good example of a company with three distinct types of users—buyers, sellers, and developers—all of whom require documentation. Take a look at **ebay.com** and browse the extensive and well-organized help that is targeted to these different user types.

What if you're expected to write a manual for a programmer, or a hardware engineer, or a network administrator? How can you write documentation for someone who knows more than you will probably ever know about technology?

Don't be intimidated—technical writers have been doing this for years. Go to the person in your own company who most closely resembles to the person for whom you are trying to write—the developer or developers who are creating the product or managing the system. They not only will tell you what one of their peers would expect to see in the documentation, they will likely help you to create that content.

Daunted by dealing with the experts? Chapter 14, "Gathering Information," explains more about how to acquire technical information from those who know and how to assemble it into useful—and usable—documentation.

Asking the Right Questions

Writing for any user involves task analysis, as you determine what tasks your users need to perform, and what steps enable them to complete those tasks. You may create a *persona* (a fictitious version of a user type), if only in your head, that describes your target user. Give the target user a name and a face so you have a concrete mental picture of the person for whom you are writing.

What is the user's goal?

"What is the user's goal?" should be your main question. Once you answer it, you know where to go from there. For example, is your user running reports with your software? She will need to know how to use the software to succeed with her standard workflow. She may also need to know how to perform additional work tasks and other user interface features.

Is your user's main goal to perform or monitor a daily backup of data on multiple servers? He will need different and much more detailed information than the person running reports, because he is responsible for data that is critical to the business. Perhaps to reach his goal, he needs conceptual theory to learn how your backup solution works before he even starts. Then he needs explicit steps to help him configure the servers. He also needs operational and troubleshooting information.

User experience (UX) designers keep the *persona* in mind as they design and refine the product. Personas are an important part of user-centered design and they can be helpful to a tech writer too as you contemplate who your user is.

Introduced by user experience leader Alan Cooper in his book, *The Inmates are Running the Asylum: Why High-Tech Products Drive Us Crazy and How to Restore the Sanity,* personas are made-up characters designed to represent the target users of a given product. Each one is usually given a name and picture, and a detailed write-up that explains the persona's usage patterns, preferences, goals, skills, and everything a that affects this persona's interaction with the product.

The more you learn about what your users need to do and the steps it takes to accomplish that goal, the better your documentation will be.

How often will the user refer to the documentation?

How often do you expect the user to refer to the documentation? Continually? Daily? Only occasionally? Will the documentation perform a time-critical task such as teaching the user how to quickly master a product's basic features and functions? Or will it be a reference work that provides information beyond the user's basic tasks? Knowing how someone uses documentation helps you decide the kind of information to provide and also how to provide it to them—

TRUE STORIES

Jay is responsible for software and firmware release notes for a highly technical piece of management software used in a specialized field. He received complaints about the release notes, but the complaints were so vague, he wasn't sure what to do to improve the problem. He added more content and explanation to them and had them reviewed by a range of people within his own company. When the complaints continued, he talked to the account representatives and product managers and asked them to speak directly to the customers to learn what the issues were. His attempts helped, but customers continued to ask for more improvements.

Finally, Jay asked for an opportunity to speak directly to the customers and found out that the problems were not as complex as he had feared. He learned there were actually two types of users consuming these release notes—one of them very experienced in the specialized field but not at all knowledgeable in the technical aspects of the firmware and software. The other user was part of a testing group that was outsourced by the customer. Jay revised his format to allow the nontechnical users to understand at a glance what was meant for them and what was meant for the testers. He included the information the customer asked for, revised language where needed, and made other changes. Ultimately, customers acknowledged these improvements in the next customer survey.

Jay's success is a huge win for his company because every complaint from a customer costs time—and money.

whether it's a downloadable PDF, searchable online help (sometimes known as user assistance), or a printed piece.

What problems might the user encounter?

Any predictable or known problems you can identify early on can become a basis for user guide content as well as fodder for troubleshooting content. Many users do not refer to the documentation at all until they have trouble.

How technical is the user?

The user's technical level of expertise determines a lot about how and what you write. Do you need to explain things in a lot of detail, or can you make assumptions about their knowledge?

Is English the user's native language?

Some documentation is published in English and distributed worldwide to all countries, with the assumption the user can read and understand English. Other documentation is translated into different languages.

Whether your documentation is being translated or not, if you know you are writing for a global audience, make sure you avoid all Americanisms, colloquial terms, excessive words, and confusing language. Chapter 21, "Gaining a Global Perspective: Localization and Translation," gives some advice about writing for an international audience.

Is this user an important customer?

It's important to keep in mind the business needs of your own organization. Although it would be nice to create documentation for everyone, the reality is that budgets are often tight and there often are not enough technical writers to fulfill the demand within a company. Companies must prioritize documentation deliverables to stay within budget. Your responsibility as a writer may be to ensure that the end user documentation has top priority. Or you may be asked to focus on customized documentation for a key customer. Or it may be decided that you will not create documentation for internal customers. Whatever the decisions are, be aware of them so you know how to prioritize your time.

Different Users Have Different Needs

Let's look at a fictitious product that involves different users with different needs. Let's say that your company, EPCS Networks, develops and sells a solution called EnterPrize Cloud Storage, which helps large businesses back up data. The company's product consists of many different software applications and some hardware components.

EnterPrize Cloud Storage will pass through the hands of many different types of people and the documentation will be used by many types of people within the customer organization and your organization as well. You identify the people who are likely to need documentation by talking to Product Management, Customer Support, Sales, and Engineering and come up with this list.

- **The potential customers**. These are people targeted by the sales staff in your company. They can be anyone from a technical staff person to a business executive who makes buying decisions.
- **The end users** who use the software or hardware for the tasks it is meant to perform.
- **The installers** who install the software on your customer's machines.
- **The Operations people** who manage the data center.
- **The developers** who use application programming interfaces (APIs) to customize your software to work with the customer company's business applications.
- **Trainers** who teach customers how to use the product.
- **The Customer Support representatives** in your company who help customers with their problems.

That's a lot of users, and that's only for the company's one core product. The company is also talking about doing a consumer product for home users who want to back up their own data, a disaster recovery solution for enterprise customers, and customized solutions for several large and important customers.

In the rest of this chapter, let's look at all of those user types and talk about what their goals typically are. All of the documentation mentioned for each user will then be discussed in Chapter 8, "The Deliverables."

The Potential Customer

There are probably many salespeople in your organization and they most assuredly use documentation to help sell your company's product or put together a proposal. The sales people aren't exactly the users of your documentation in the traditional sense, but they certainly do put your documentation *to use*. (One good thing about salespeople within your company is that they won't hesitate to ask you for what they want.)

Sales representatives are looking for documentation that is informational, and explains what the product does in a way that can convince the potential customers that it will solve their needs. Documentation for this type of user includes white papers and user guides.

Just because someone understands his business does not mean he will magically understand how to use your product. Often assumptions are made that as long as the features are in there that enable users to do every single thing they may want or need to do, that your only job is to explain the features and functionality and let the users figure it out.

This rarely works, because the users need to know the *workflow*: the series of steps that enable them to perform relevant tasks. Writing for these workflows is much more difficult than simply explaining all the functionality on the menus.

There are also a lot of documents that go into the sales process, such as requests for proposals (RFPs). Those documents are normally handled by departments like Business Development or Project Management, but they often contain portions of technical documentation. And sometimes technical writers are called upon to create those documents. If you get the opportunity, take it—it's good to be involved in something that helps win business for the company.

The End User

The end user is the ultimate consumer of a product, the person for whom the product is designed. As a technical writer, you are the end user of a product such as Adobe FrameMaker or Microsoft Word. The documentation you refer to for those products teaches you how to create and save a document, insert a hyperlink, run a spell-checker, and perform other tasks so you can do your work.

The end user's goal is to learn how to use the product as it is designed. This means they are using it to perform a job function (like using Adobe FrameMaker), or an entertainment function (like playing a game or reading a book on an e-reader) or for personal use (like a phone or camera or any consumer device). End users want information that is immediately available and solves their problem with the least amount of interference. In its simplest form, this may be a single word on the display of a phone app. Yes, technical writers often write those messages, too.

The novice

A novice is your new user, the person who knows nothing about your product and is trying to learn enough about it to put it into use. You must always provide documentation for the novice end user. Novices know nothing about your product, although they may know a lot about the business for which this product is targeted. In writing documentation for the users of the EnterPrize Cloud Storage solution, you must assume they know nothing about how to use the solution, but they know a lot about their data and their servers.

If you were writing documentation for financial software, for example, you should assume that the users have never seen your software before, but they know all about accounting or being a CFO or handling their budgets. It then becomes important for *you* to learn something about the novice user's purpose for using this product.

Novices need something short and simple with which they can get started quickly. Their goal is to perform the tasks they need to do their jobs or to start enjoying their new "toy," but they are not necessarily interested in reference material or advanced tasks.

Examples of documentation for novices include user guides, quick-starts that help them jump straight into performing a product's key task (think making a call with a phone or taking a picture with a camera), and tutorials that teach them how to get started. This user also needs many forms of help, which tells them how to accomplish what they are trying to do, and on-screen help, which tells them exactly what to do *now* and in *this place*. There's more about help in Chapter 12, "You Want it *How?*"

The expert user

The expert user knows how to use the product and needs information about how to do even more with it. Unlike the novice, the expert user does not want to read through detailed, repetitive procedures with every step spelled out.

Expert users have typically been using the product for a long time and will refer to the documentation when there is an error scenario or if they want to perform a new task and can't figure out how to do it. Their goal is usually to figure out how to do something with the product that is not one of its basic, most common functions.

The expert user wants searchable or *context-sensitive* help (in other words, help that is available on the spot. See Chapter 12, "You Want it *How?*" for information about context-sensitive help). The expert user has a task in mind and wants to accomplish it as quickly as possible, and wants to find out how to do it without reading pages of text or stepping through procedures with which he's already familiar. The expert user will refer to user documentation as well, but definitely does not want to plow through a lot of beginner material.

The expert user may want a quick-reference card to be used as a reminder, whereas the novice user needs the same steps delivered in step-by-step detail. An expert user will also depend on release notes with new versions of software. Since the expert user knows the product well, it's important to see release notes to learn about the changes in a new version. I talk more about release notes in Chapter 8, "The Deliverables."

Examples of documentation for the expert user include user documentation in the form of online help, release notes, troubleshooting, and reference.

The consumer

Consumer products like off-the-shelf software, electronics, appliances, and other devices have their own special documentation needs. Because a consumer has so many choices in the marketplace, a company must make a special effort to cause a customer to pick that product off the shelf or off the site of an online retailer, keep it, and remember that brand when purchasing again. Documentation can play a part in customer satisfaction.

Out-of-the-box experience (OOBE) refers to the first impressions a user has with a product when opening its packaging and starting to use it. Apple is one company that seems to understand the importance of the out-of-the-box experience.

The ease of use and the "wow factor" of the packaging all contribute to drive customer loyalty. Many users cite hard-copy documentation and attractive document design as important factors in a positive OOBE.

There are many documentation decisions that go into what might seem like even the simplest of products, such as a consumer electronics product that is sold on the shelves of a big-box store. What are the things you need to take into consideration to document a product like this?

The store—and your company—do not want customers to return the merchandise because they couldn't figure out how to set it up. Often the product is packaged with a printed booklet in the box to enable the customer to set it up and start using it immediately.

Your company doesn't want to scare customers with long, complicated instructions that make the product seem difficult, so the printed guide should be short and easy—a quick-start guide. Often, this is supported by PDF or web-page documentation available from the company website.

With any consumer product, the out-of-the-box experience is important—everything from packaging to starting up and using the product quickly and easily must be a positive experience for the customer. An attractive, easy-to-follow user's manual is part of that experience.

And lastly, manufacturers of consumer electronics are in a constant battle with the competition to keep prices low. While the quick-start guide should be clear and attractive, it must also be relatively inexpensive to print and produce. This might mean limited colors or sizes, and limited page count.

But users usually need more than a quick-start can give them. You may decide to also provide solutions on a knowledge base and a monitored community forum where users can share ideas. You may be asked to write content to get the message about this product out to users who follow your company on Twitter or Facebook. You may produce videos to show how to set up the product. And you are likely to also provide PDF documentation available from your corporate website.

TRUE STORIES

Nan's story: Once I was at a customer site and an in-house consultant from my company was helping the customer install our enterprise product using the installation guide that I wrote.

The consultant was looking for a URL on a screen shot, but it wasn't there because I cropped it out. I had prioritized the part of the window that the user would see as they followed the instructions, and I put that screen shot there to orient the user. But the screen shot was useless. Now when I do include screen shots, I am more discerning. I carefully consider whether the screen shot is necessary and, if so, why. I always ask myself this question: "What does this screen shot bring to the table for someone who just wants to complete a task and move on?"

The Installer

Installation documentation must be comprehensive, clear, and help the installer avoid problems. If you're like most of us, I'm sure that you have your own frustrating stories of trying to assemble something or install software using instructions that didn't make sense. Put yourself in the shoes of the user and make sure the documentation includes everything that is needed.

The goal of this user is very simple. It is to install software or a service, or to assemble something so it will work and work correctly. Correct installation may include the need to configure and to troubleshoot problems as they occur.

If you are documenting software, it is usually easy to find someone within your own company who can act as the installer. People in QA, Engineering, and

Operations install the product all the time. If this is the case, you have the ideal combination—your internal team is your subject matter expert and user all in one. Take advantage.

Documentation for the installer includes wizards, standard installation instructions and guides, quick-reference steps, upgrade instructions and guides, configuration material, and quick-start installation cards or posters.

Operations

Operations people can be anyone from the IT department, to people who run the data center, to people who run the network. Their goal is to keep the company running day to day, or to manage services for customers. If your software or hardware is going to be part of what they are managing daily, they need to know how to keep it from going down.

Documentation for this type of user includes operations guides, networking guides, troubleshooting documentation, *run books* (a collection of procedures used to run system operations), reference documentation, and system administration documentation.

Developers

Developers, engineers, programmers, designers—no matter what you call them, they are the people responsible for building the technology. Developers within your own organization will often write documentation for themselves and their peers, perhaps on a *wiki* site (a collaborative website that can be modified quickly by many users), perhaps in email, but in many companies you may be expected to provide developer documentation for your internal team.

You may also be asked to write developer documentation for customers. This will often be API documentation. Their goal will be to learn how to integrate your product with their business systems. Developer documentation in your own organization may be internal documents, like design documents, use cases, or technical requirements documents. Refer to "Internal Documents" on page 100 for information on these.

Trainers and Trainees

Trainers are responsible for teaching customers how to use the product, which means that both they and their students are the users of the documentation you create. The trainer's goal is to learn enough about the product to be able to teach her trainees what *they* need to know. Typically, a trainer uses a slide deck with very high-level information and fills in the information with his own specialized knowledge.

It is very important for you, the person who provides customer documentation, to have a good relationship with the trainers in your company. Trainers are great testers for your documentation, as they learn the product. They are also in a good position to tell customers about the documentation, often distributing product documentation during training sessions. Most important, one of the best ways for you to learn how to improve the documentation is by sitting in on a training session.

There are also training materials you may be asked to provide that aren't designed for a class but rather are materials designed for self-training, such as tutorials and interactive pieces (often called computer-based training, or CBT).

Relying on others to check the accuracy of what you write is critical, but there's nothing like watching real users to see how they handle your company's products. Jump at any and every chance to use the product, take training along with the customers, and meet your users. The insights you'll gain from even a short interaction will be invaluable.

Documentation for the trainer and the trainer's students includes training slide decks, handouts, and any guides, manuals, and quick-references they may provide to their trainees. If you do not write the training materials, make sure you work regularly with the training staff to make sure their documentation and yours are in sync.

Customer Support Representative

Customer Support representatives participate in calls, online chat, or email with customers and solve their problems. Because of the time (and therefore cost) involved in this direct contact, many companies try to provide as much "self-help" as possible to customers on the company website. This is often in the form of an FAQ (Frequently Asked Questions) and a *knowledge base*, a centralized repository of information which the customers access by searching or typing questions.

Build a strong relationship with someone in Customer Support. Every problem a customer calls about should be considered for inclusion in documentation. If it's already documented, ask yourself why the customer doesn't know about it. Then see what you can do to correct that situation.

Besides the information provided to the customers, the reps themselves need the most up-to-date documentation so they can either provide it to the customers or refer to it themselves to answer the customers' questions. The support rep's goal is to be able to solve any problems the customers have. They need to be able to quickly access the information that will help them do it.

Customer Support is another department with which you want to have a good relationship. Support reps know as much as anyone in the company about the problems the customers have and they are an invaluable source of information. They are also a very good internal customer for your documentation.

One Document, Many Users

Technical writers often must create one document that serves the needs of different user needs. This may be because of budget reasons. Or sometimes it's because the company doesn't want to scare users by giving them too many different guides—it makes the product appear to be too complicated.

This can mean supporting two totally different functions, like a combination user guide and installation guide, or it could mean providing information for customers that range from inexperienced and nontechnical, to highly experienced and highly technical. Merging functions in one document requires some planning and creativity, since you don't want to discourage either user type. A beginner may need detailed steps, while the experienced users want only to have their questions answered, or to read conceptual information that can help them solve their own problems.

Sometimes, if you listen to customer complaints, you'd think all of the documentation coming out of your group is a failure. Remember that customers rarely call to praise, and the messages you'll hear from customer support are typically going to be from unhappy customers who tried to do something and discovered that the documentation was incomplete or incorrect.

Accept the complaints with a smile and make the corrections immediately. Be happy that your customers are using the documentation!

One way to provide information for both is to give a quick-start at the beginning of the document. The quick-start is a single page that contains brief steps, with references to more information. If you know the majority of your users are more technical or more experienced, you can write the steps for them, with a clearly marked box or note below each step for the beginners. I've even seen documents that contain two separate sets of instructions for the same procedures, each designed for a different user type.

So *That's* the Way They Do It...

Try to learn how the customer really uses the product. If you write instructions that assume one thing while your users are actually doing something completely different, you are adding difficulty to your customers' lives as they figure out how to modify your instructions to fit their own reality.

As well, find out how the customer uses the documentation. Are they outside doing a field installation and trying to read a PDF on their laptops? Are they unable to read your help because they use the Google Chrome browser and you only tested on Internet Explorer? Are they trying to find out why they have no phone connection but the help on their mobile devices requires Internet access?

Meeting the Users

Audience analysis—even the briefest sort—can help you tremendously in getting started. If your company has a User Experience or Usability group, align yourself closely with them. Content is a key component of the user experience, and it's important for you to do your part in making the product usable by providing the right content at the right time. If your company's User Experience group has a usability lab where user testing is conducted, ask to observe the user tests so you can note where the participants have problems.

Attend training classes, whether internal or external. Doing so helps you in two ways—you get much more knowledge and depth about the product you're writing about, and you see what questions and problems people have. If documentation is used during the course, it also lets you see how people use the documentation. If you are able to attend a session with external customers, meet those customers and talk about what they would like to see in the documentation. Find out if they even know that documentation exists. (If they don't know—which happens—that will be your first problem to resolve!)

If your company does not offer training, try to meet your customers through trade shows, telephone interviews, or survey questions. There are many inexpensive online survey tools available, and they provide a good way to get answers to some of your questions.

Take advantage of social media outlets to "meet" and collaborate with your users. This kind of two-way street can be a great opportunity to be involved with the people who care about your product. Chapter 12, "You Want it How?" talks more about how social media and technical writing can come together for everyone's benefit.

Not Meeting the Users

In many companies, you won't have the time, budget, or authority to meet the people who use your product or your documentation. There are still some tricks you can borrow from the user experience field to define your user.

- **Find out who your company thinks the target user is.** The *product manager* (the person responsible for selecting and determining the features of a product and overseeing it through development) should have a very good sense of who the target customers are and what they need to know. Product

Management has likely done market research to learn what customers will look for in this product. Read the product requirements document or marketing requirements document to learn what customers expect.

- **Become your own best customer.** If your company's product is a consumer product you might use yourself, great! Make notes of the things you want to know and the things that confuse you. If you want to know these things, other users will, too. Take lots of notes before you become too familiar with the product; too much familiarity often means you skip information that's important to novices. Figure out what your goal is with this product and make sure you cover everything that enables you to achieve that goal.
- **Find a surrogate user.** If you can't meet the customer, find someone who has a similar role and similar characteristics. Think about the personas discussed earlier and seek out an individual who best matches the target user.

Section 508 Compliance

Section 508, an amendment to the United States Workforce Rehabilitation Act of 1973, is a federal law mandating that all electronic and information technology developed, procured, maintained, or used by the federal government be accessible to people with disabilities. Technology is deemed to be "accessible" if it can be used as effectively by people with disabilities as by those without.

One way to make PDF documentation compliant for an Assistive Technology (AT) device is to make sure you always use tags and structure correctly in your source files. Changing character formats by applying bold or size changes to existing fonts does not work well with AT tools. Instead, be consistent in applying predefined tags like "Heading2" or "ChapterTitle" to paragraphs.

Your users not only have different levels of technical expertise and different business needs, they often have different accessibility requirements. Documentation, from PDF to HTML, should be Section 508-compliant. Go to **section508.gov** to learn more about what's required.

It is a good idea to check all of your electronic media for accessibility, even if the US government is not currently in your customer base. Not only do you want your customers to be able to read and use your documentation, but also if you do try for a government contract, it can be a lot of work to retrofit everything. (I have been part of a couple of last-minute scrambles to try to win a government contract. Proving compliance, or adding it after the fact, is a sizeable task.)

Chapter 8

The Deliverables

The stuff you're hired to produce, and some thoughts on who might use and read it.

What's in this chapter

- The types of technical documentation you'll be called on to write
- Why task-based documentation can help
- What to call a bug
- Using the same content in many places

There's no recipe to tell you exactly what the a specific documentation deliverable should be made up of. Instead, as you go through the processes described in this book—from learning what your users need to developing the content—you'll find yourself building a complete set of product documentation. It's an iterative process: your first deliverable may not be as complete as you would like it to be, but if the product is successful, there are always follow-up releases and further opportunities for you to improve.

Does that mean there is no way to figure out what information needs to be included in any given type of guide or online help or tutorial? Not at all.

You will rarely be working in a vacuum. When you join a company (and I'll talk more about this in Chapter 13, "Getting Started"), you'll typically find that there is already documentation of the type you are to produce. There may even be strict department guidelines about what sections belong in these documents and what these sections should cover. In that case, once you understand for whom you are writing the documentation, you can take what already exists and copy and modify it for your project.

This chapter does not focus so much on the method of creating or presenting the types of documentation—that will be up to you, your users, and your company's business needs. Chapter 12, "You Want it *How?*" discusses the tools you can use to create any kind of documentation in any format you want.

Building the Documentation Family

It is a good idea to start out by defining a few types of documentation deliverables and build around them. As you develop more and more documentation as it's needed, you can add more to a product's documentation library. For example, you might decide that each new product gets an installation guide, online help, and release note. As the documentation becomes more mature, you may add more.

You can make your life easier by making decisions about what types of information go into what documents,. then avoid copying information from one document into another. Tempting as it may be to assist your user by including key information in every place where you think it can be helpful, it can cause problems with maintenance and consistency.

It's a sure thing that you'll update information in one place but not another. It's embarrassing when you discover that one of your documents contains outdated information while it is correct in another!

Even when the content is essentially correct in both places, when one person tweaks the content without realizing it exists somewhere else, after a while, there are two or more same-but-different passages in multiple documents. The user is confused, wondering if the different sections mean the same thing, or instead have slightly different meanings.

A way to avoid those problems is by applying single-sourcing and content reuse principles.

Combining It All: Single-Sourcing and Content Reuse

No matter what your final output, this is a good time to start thinking about the fact that you will probably plan to use a great deal of your content in multiple output forms.

Single-sourcing refers to the development of content with the intent of producing all or parts of it in multiple formats. Instead of copying content into different files and formats and maintaining it in more than one place, the content exists in one place only. Typically, this means that you have many files stored in one location and you use some or all of them in each individual deliverable.

Content reuse refers to the management of content by breaking it into small enough components, or topics, so that each topic can be used in the appropriate place. *Topic-oriented writing* is writing that is intended for reuse. Each topic is a unit of information that stands alone, or can be mixed and match with other units. If you're accustomed to writing in a linear, narrative style, topic-oriented writing may require some effort to learn.

Single-sourcing can be as simple as creating a single file of content that you then generate for print, PDF, and HTML. Or it can involve the creation of many small modular topics and conditional content files that are tagged, mixed, and matched to build different types of output. In all cases, changes are made only to one file—the source file.

For example, the help and user guide for an application are likely to share most of their content. In fact, if you are creating a user guide, online help, a data sheet, Facebook content, and help for a mobile device for the same product, they are all very likely to share some content.

> **INSIDERS KNOW**
>
> It can be OK to repeat content when you have a single-sourcing plan. Using a single-sourcing plan or text insets in FrameMaker allow you to place the same content in different output. When a change is made, it's made to the original source, not to a copy.
>
> But don't let single-sourcing be an excuse for repeating the same content in too many places. Instead, think about the best way to present content, and make sure the users are able to find the right content when they need it. When you do reuse content, do it as part of a well-thought-out plan and you should have a good reason for repeating content when you do...users searching for information don't want to find the same content everywhere they look.

Working from a Single Source

Let's assume you start by writing content for EnterPrize Cloud Storage in your help authoring tool. Your first deliverable will be a help system for your company's web application. Next, you use conditional text features to modify the content. With a push of a button, the content is output in book form and published as a PDF. Next, you use a different set of conditions to generate two sets of help for customers who receive a customized product. Lastly, you add some new content, apply conditions again, and produce a data sheet. One source of content, tagged appropriately, produces output for five different deliverables.

See Chapter 12, "You Want it *How?*" for more about planning and implementing single-sourcing and content reuse.

Determining the Content Approach

There are many decisions to make about how you present the documentation content. These decisions depend on what your users need to know and how they will use the documentation. They may also be based on how much time you have to work on the project and how many different deliverables are produced from the same content.

The content might appear in many different forms, as we will learn in Chapter 12, "You Want it *How?*" It can be unique, appearing only once, or it may be reused across various deliverables, targeted for different users or different platforms or different customers.

In addition to those decisions, you need to decide whether your documentation is task-based or reference-based. Choosing one of these forms depends not only on what the user needs, but also on how much time you have to work on the project and how much content reuse is involved.

You can get a jump-start on creating task-based documentation if the developers in your organization create *use cases*. A use case is a written description of an interaction between an outside "actor" and the system to accomplish a specified goal. The actor can be a specific type of user, or even an outside system. The use case captures who does what and why.

Use cases are helpful in building software because the resulting product presumably is built based on what the user needs to do rather than a bunch of features that might or might not tie together. As a technical writer, you could not only find yourself benefiting from use cases, but you might be asked to write them as well.

Task-Based: What the User Does

Task-based documentation is about what the *user* does, with instructions on how to operate the product to accomplish the tasks the user performs. This could involve moving back and forth between different screens, different menus, and perhaps combining a number of shorter processes to accomplish the goal.

As you develop task-based documentation, make sure you clearly define the starting point and the stopping point for each task. In other words, what is the first step the user does to start the procedure? And what is the final step that indicates success? Use a consistent format to present each task—a description of the task and its purpose, a lead-in heading, and a set of numbered steps are typical.

As you write these tasks, you may realize that the software or hardware has functionality and features that you never mention at all. Many products have a lot more features than are commonly used—think of how many things your phone and camera can do that you've never even tried.

Some products have legacy functionality that isn't used in any workflow. Those little-used functions can be saved for a reference manual or a more detailed version of the user guide. (You *will* have to document them somewhere, because customers of older versions who used those functions will want to know where they are in this version.)

Reference-Based: What the Product Does

In the reference-based approach, the focus is on what the *product* does. What does it look like, what does each menu, item, and button do, what does a command mean, what information goes into a text field? It's up to the users to decide how to put all these components together to accomplish their goals. Because of that, it's not always the right way to provide user information

But reference-based documentation has its place. There are many types of documentation in which this is the only way to go: glossaries, command references, descriptive documentation of all types, and release notes, for example. If what your user is looking for is straight information, that's what you should provide.

It also takes less time to develop than task-based documentation, because there is less thinking and planning. When you are short on time, or have the budget to produce only a single piece of documentation for your product, you may choose to provide reference-based user documentation. You can plan to go back later and expand it to a task-based approach.

To create reference-based user documentation, go through the application's or device's user interface and describe every single component accurately and consistently. That means every screen, every entry field, every menu option, every button, every dialog. If it's a command reference, that means every command. If hardware, every component.

You say to the user, "Here is each part, and here is what that part does." You don't have to worry—at least not yet—about how the user puts all of these functions together to accomplish something.

Delivering the Deliverables

In Chapter 7, "It's All About Audience,"we learned something about the users for whom you'll be writing and discussed the types of documentation those user types may need to fulfill their goals. Now let's go into more detail about the types of documentation you will be delivering and what is involved in developing them.

Presentation *method* is still up for grabs at this point; any of these types of documentation, for example, can be presented in many ways. We will go into that more in Chapter 12, "You Want it *How?*"

End User Documentation

End user documentation explains how to use the features of the product. Designed for the people who actually uses the product, the documentation helps the users to meet their business or personal goals.

End user documentation is so important, it often comes in more than one format. A product may have online help in the software, a short printed user guide in a box, and a long user guide in PDF form on the corporate website. There might also be a tutorial, an FAQ, and more.

Many tech writers enjoy writing user documentation because they can easily put themselves in the user's shoes. However, don't make the mistake of thinking that you can write good user documentation by waiting until the product is fully developed, sitting down at your computer, and writing about exactly what you see. What does this do that the users can't do themselves?

The way you add value as a technical writer is to understand what the user needs to do and understand how to use the product to accomplish this. Since you have access to the developers, you are also in a position to explain functionality that isn't easy to see or find, and shortcuts for doing important tasks.

User guides

A user guide can include anything and everything, including installation information. If your organization delivers only one document with a product, it is likely to be a user guide. "Guide" doesn't always mean a book, either. Your user guide may be a print document, a book in PDF form, or a set of help files.

In general, the purpose of the user guide is to explain how to use the product to accomplish the user's business or personal goals, but in fact its most important function is to get the user up and running immediately

Did you ever notice the printed user guides that come with cameras? Often, they include only one task— how to take a picture. The manufacturer knows that unless the customer is able to immediately do the thing a camera is meant to do, nothing else matters.

It's usually best to write task-based documentation for users. When a user can't figure something out, he doesn't ask, "Gee, I wonder what the XYZ feature on menu A does?" He is more likely to ask something like, "How do I save a file?" or "How do I connect my DVD player to my television?"

At some point, you can expand the task-based documentation by adding reference information about the menus and each item on the user interface. You can also add more steps for more complicated tasks. Think of the expert user discussed in Chapter 7, "It's All About Audience." Expert users already know how to do the basics but will be looking for information on how to do complicated tasks, and you need to write for them, too.

Error messages

Error messages are hugely important to any user at any level of experience. When something goes wrong, it's not good enough to tell the user she is a victim of error #403FGGG. Turn the error message into first-response help by explaining what the problem is, and what she should do to attempt to resolve it. If you are not able to work with development to insert these descriptions into the error messages themselves, you can create error documentation that lists each error code along with a description and a resolution.

It's a good idea to develop a standard for error messages. If there is none at your company, volunteer to drive the effort. Bring together Engineering, Product Management, and Technical Publications to come to an agreement on how error messages should be worded.

Error messages are often overlooked in any development organization; they are not glamorous and they often are done in a hurry by the developer working on the software. When a technical writer is responsible for error messages, they improve, if only because the tech writer has to dig to ask the question about what the error means and then to write the message in a way that makes sense.

Tutorials

Often it is better to *show* customers what to do rather than write a long procedure that tries to *tell* them what to do. Tutorials should teach a user to do something for the first time.

A tutorial is likely to be used once, and not by every user. However, the people who do use it will be grateful for its help. (And may reject your product if the tutorial does not help.)

Tutorials can be a lot of work, especially if you do an interactive tutorial. Be prepared to spend a lot of time going over the layout, testing the visuals, and changing the timing. I don't recommend that you volunteer to provide a tutorial unless you know you'll have plenty of time to create it, and even then, I suggest that you plan these only for very important, highly visible products that require user hand-holding. See "Multimedia" on page 158 for more information.

Tutorials can be done in a cost-effective manner with a recording program like TechSmith Camtasia to record short clips of on-screen activity as a supplement to existing documentation. These clips are often used on knowledge base or customer support sites to show customers how something is done.

Help Content

"Help" is a broad term. It can be said that an entire collection of documentation is user help, and that's true. However, most technical writers reserve the term for the task-oriented modules that come up when a user clicks a Help link or icon on a software application. Because it's a presentation mode, I devote time to discussing it in Chapter 12, "You Want it *How?*"

But it is so important to an end user, I need to introduce it now. Often known as "online help" and sometimes as "user assistance," it is often produced with a help-authoring tool (HAT) that enables linking, searching, and indexing of various topics as well as generating the content into a file format that works with different platforms.

Successful help requires a completely different structure from that of a book. Books are typically written with the expectation that a user can start at the beginning and proceed to the end in a sequence of events that result in a complete story. A book can include conceptual information and supplements to let the user know all about different aspects of the product.

Help, on the other hand, is meant to be accessed because the user wants to solve a problem right now. When the user clicks a Help icon, she wants to know how to install a mobile application, or how to fill out the forms on a web page, or what certain terminology means. Once the problem is solved, if the user wants more related information, there should be a way to get that, easily.

Your user wants easily available answers for the most common and important activities. This could mean that you provide the content right there on the web page so the help is right in the user's face. It could mean that a mobile app is so intuitive that any explanatory content is right there on-screen and there is no need for additional help modules. It could mean that you provide content as a mouseover or pop-up, so the help appears when the user moves her mouse or clicks a help icon. Or it could mean that the help appears as an FAQ or on its own set of web pages. However you present help, it's your job to determine what the user needs to know and how to make it most easily available.

Chapter 12, "You Want it *How?*" provides information about how to plan and create help.

Installation Documentation

Anyone from your grandmother to the network administrator of a software company, may find herself in the installer role...it all depends on the product. Installation can involve anything from clicking a couple of buttons on a wizard to configuring numerous servers and installing many different software packages in specific order. In all cases, you must make sure the information is correct, complete, and presented in the right order.

Installation documentation is needed, in the best of all possible worlds, once. This means it's usually best to keep it completely separate from the user documentation, which is expected to be used regularly. The installation documentation should be available before the actual product is installed, which means it will be distributed in a format that makes this possible:

- A hard-copy printed piece
- A PDF or web content posted on the company website or extranet
- A wizard or setup assistant that leads a user through a series of well-defined steps
- A web page or web help

> **INSIDERS KNOW**
>
> Installation documentation must come in a form that allows the user to have immediate access to it. This can be anything from a hard copy manual included in the box, poster, on-screen wizard, to a PDF that the user downloads from a company's website.
>
> Electronic devices normally include installation wizards and sometimes additional documentation that is actually stored on the device itself. If this is how your company provides documentation, make sure users have a way to get back to it if they don't download it while they are going through the installation wizard.

When planning installation documentation, think about all of the following:

- **Prerequisites.** Let the user know everything that must be in place before the installation can start. Does the installation require a specific version of Oracle software? Does it require a certain amount of disk space or memory before the user can begin? A digital certificate? There's nothing more frustrating than starting to do an installation and only then discovering that you can't continue because there is a whole list of things you needed to do first. Especially if those prerequisites are activities that require a lot of time or cost money.
- **Context.** Tell the user where to start. Does he unzip a package of files onto a specific server? Does he log in as a specific user? Put a DVD in a drive?

- **Dependencies.** What else can affect this installation? Is there a specific step the installer must take if two applications are installed on the same (or different) machines? Does installation of one component require the installer to then do something to another component?
- **Troubleshooting**. What does the user do to resolve a problem if something goes wrong during the installation?
- **Upgrade.** What steps does the user take if he is upgrading to a later version of software rather than performing a fresh installation?

These are all issues that must be considered for even the simplest installation.

It's very important for you to do the installations yourself to write accurate instructions. If your organization doesn't seem to think that's necessary, it's because they don't have experience with tech writers who actually write their own documentation.

Release Notes

Release notes accompany products when, as the name implies, there is a new release, or version. Release notes provide the user with information that does not belong in or did not make it into the regular customer documentation.

For example, you will typically explain in the release note what's new in this release. That wouldn't be appropriate in the user guide, because the user guide is written for both first-time users and returning users. It should not assume the reader knows anything about a previous version. (Also, user guides are frequently not *version-dependent;* that is, the same user guide can be used across many software versions assuming the functionality and user interface remain essentially the same. This means that when the same user guide is used throughout many releases, all product changes are documented in release notes.)

Release notes also contain information about bugs or workarounds. This material also does not belong in the main document, because these bugs will presumably be fixed.

Release notes are delivered in a variety of mechanisms, sometimes as *ASCII* text files, that is, plain unformatted text that can be read on any operating system

Where did the term "bug" come from? Nobody's sure, but it was already in use in 1947 when a moth got between two electrical contacts and shorted them in the huge Harvard Mark II computer developed by Howard Aiken and the team backed by IBM.

A technician's log entry noted that it was the "first actual case of a bug being found" and he taped the proof—the moth itself—into the log book. Now that's what I call documenting!

(often called READMEs). Because they are written at the very last minute, as bugs are fixed and QA testing is wrapping up, they need to be done in a format that can be easily generated and distributed to the customer.

Whatever format you choose, make sure your release notes are clearly written. If you don't understand the bugs you're describing, your readers won't either. Make the notes as short as possible while still including all the information the user needs to see. Normally, customers want to scan release notes quickly to see what affects them.

Many different types of users depend on release notes. These documents can be difficult, because you are writing for everyone from the naive end user to the highly technical user, as well as for executives who want to know whether to upgrade to the next version. You have to use a style and voice that speaks to them all.

Release notes can include just about any type of information the development team or product manager thinks should go in, such as the following items:

- New features in this release
- New documentation for this release
- Bugs that were fixed for this release. (Don't call them bugs, though; you'll have to give them a more positive spin. Many companies use the more neutral "issues.")
- Bugs that you know exist but haven't been fixed yet, along with workarounds for the problem
- Upgrade information the customers need to know for this release
- Tech support information
- A list of which files and directories will change after the new version is installed

Workarounds are solutions to bugs or other software problems, enabling a user to "work around" the problem until it is fixed. Often those issues never become fixed, because if a workaround exists, the team may devote its time to fixing problems that don't have workarounds.

Troubleshooting Documents

Troubleshooting documentation contains descriptions of problems the user might encounter, with information about how to solve those problems. Everybody wants them, but they are often the last piece of documentation to be produced, because they are just that difficult.

The writer sometimes needs to be a bit of a psychic to guess what problems might occur, and a detective to run around and gather up information on these likely problems The biggest difficulty is that frequently, no one knows what the problems are until they happen. And then everyone is so busy trying to solve the problem, they certainly don't stop to write down what happened or think about letting you know about it. Because an effective troubleshooting guide requires history and experience, it's rarely possible to do a good one right away.

If you are asked to create troubleshooting material in the first release of a software product, there are a few techniques you can use to gather content. Use your own experience with the product and think about what areas may cause problems, then talk to product management and product development to find out what they think. Talk to QA, Operations, or any other group who has had experience installing and running the software and ask what problems they have encountered and how they've solved them. Review the existing bugs in the software and the workarounds provided; it might be useful to repeat this information in your troubleshooting guide.

Ask the developers if they can extract the error messages from the software so you can see what errors are expected to occur. Producing a list of the error codes and messages along with descriptions of what the users should do when they see the error is a great start for a troubleshooting document. The *escalation path*—the steps a customer takes to recover from the error, is an important part of this documentation.

Developer Documentation

As a tech writer, you won't always get to write documentation aimed at a user who resembles yourself. You are likely to also have to produce documentation for much more technical users. These are users who usually have much more extensive knowledge about the systems but need reference material or API documentation or design specifications to help them do their jobs.

If you used to be one of those people yourself, you have the perfect background to write these types of documents and are likely to be in great demand as a technical writer. But if you don't have that kind of background, you must be very careful. You must write solid, accurate material and make sure it is reviewed closely by someone who understands it.

Developer documentation is much easier to write if you can read the code; often, the developers add comments that contain a lot of explanation. As with all technical writing, the more you understand your product, the better your documentation will be.

Many developers have experience with technical writers who changed the wording of something to make it "sound" better, who changed the punctuation or capital letters in Linux commands because they didn't understand them, or who took notes that made no sense later but the writer was embarrassed to ask what they meant. The resulting documentation was then wrong, sometimes disastrously so.

Until you become knowledgable about the product and your audience, ask for help. It may be annoying to developers who don't have the time to do what they see as your work, but if you explain up front that you don't have the knowledge—yet—and you're working to learn it, that's the best you can do. In the meantime, beef up your product knowledge, read industry material, or shadow a QA person. Refer to Chapter 10, "Become Your Own Subject Matter Expert," for other ideas on learning the product.

API Documentation

An application programming interface (API) is a set of rules and specifications that software programs follow to communicate with each other. It acts as an interface between different programs. APIs are what make your Flickr photos appear on Facebook and a Google map display on a real estate web application.

APIs can be important for customers who want to use your organization's code to integrate your product into one of their existing applications. The structure and content of an API document depend on the components in the actual API itself. The API document must provide all the information necessary to enable another application to interface to your product and to perform all important tasks associated with the API. The document should list all functions and their meanings.

The best way to tackle an API document is to find out what the programmers in your company believe should go into it. These programmers are often just like the programmers in your customer company, after all. There are many examples of API documents online. Try taking a sample of what you see online as your template and review it with the developers in your organization until you come up with something they think works.

User Manual, User Guide, or User's Guide? Release Note or Release Notes? Reference Guide, Reference Manual, or Reference?

It's not important what you choose to call the different deliverables as long as you are consistent. If someone on your team has a strong preference for one of these terms, accept it, define it in your department style guide so everyone knows what the content and structure ought to be, and move on. It makes no difference to the users for whom it's intended as long as they know what information to find there.

Internal Documents

Internal documents are the documents that are used within your organization and not meant for customer consumption. If your Tech Pubs department only produces external facing documentation, you won't be called upon to write these; however, you will definitely use them. They're not something you'll find in every company. They exist in many mature companies, but many startups do not bother with them.

You might find, for example, that instead of a requirements document, there might just be a slide deck. Instead of a design document, there might be only a seasoned developer who holds all the information in his head. Perhaps there is a full set of requirements for a product, but it is not produced in any kind of standard format you're familiar with—instead, it's a series of tasks or features submitted in a bug and issue tracking tool such as JIRA. Or it's briefly described in a Microsoft PowerPoint presentation.

Tech writer Ken needed a solution to the difficulties of getting emergency release notes out when he didn't have time to draft, review, and release. He came up with the suggestion of JIRA, an issue-tracking tool used for bug-tracking and project management, to the development team.

JIRA was adopted by the engineering team as their bug-tracking tool and Ken had a collection of descriptions and comments for each bug to use as release note content. If he needs his content reviewed for any given bug, he posts a comment to JIRA and to a targeted group of reviewers who are already working on the bug.

Finding a solution that both the developers and tech writers could successfully use has enabled Ken to meet release note deadlines with the tightest turnarounds.

See **atlassian.com/JIRA** to learn more.

Requirements documents are typically written by product managers or product marketing people, and specifications are typically written by developers. However, you can make yourself become a very helpful member of your organization if you are able to write them.

Even if you don't write these documents yourself, you will surely want to use them to help write your own documentation, so make sure you know what they are and who is responsible for them. You can write quite a bit of customer-facing documentation from a requirements document, spec, or design document without ever seeing the product.

Requirements documents

Requirements documents explain what a product needs to be. A *marketing requirements document* (MRD) describes what the product must include to meet customer needs. A *product requirements document* (PRD) defines the product and the features it must have. Many companies merge these two documents into one, so keep your ears open for the acronym in use, or just ask someone if there is an MRD or PRD available for the product in question.

Not everything in the PRD appears in the product. Requirements often use a method such as MoSCoW to identify the priority of different requirements. Low priority requirements might not be implemented in the final product.

Here are the four categories of requirements in the MoSCoW method:

- **M - MUST:** Describes a requirement that must be in the product or solution.
- **S - SHOULD:** Represents a high-priority item that should be in the product or solution.
- **C - COULD:** Describes a requirement that is desirable but not necessary.
- **W - WON'T:** Describes a requirement that will not be in the product or solution.

Specifications

Specifications are documents that define what a product or application does. Specs may be written in response to a requirements document, and design documents may be written in response to a specification.

A specification can define anything from documentation to software to a piece of hardware. For example, a document specification would define how documents are named, versioned, laid out, and formatted and structured. A functional spec defines the behavior of the software in response to input. A user interface specification describes how the user interface (UI) behaves when a user interacts with it.

Design documents

A design document describes how the software works, and describes the software architecture. There are often two types of design documents: high-level design documents (HLDD) and low-level design documents (LLDD).

The HLDD is more of an abstract view of the system with less detail, and the LLDD, as its name suggests, includes enough detail for a developer to do implementation from.

Oh, and Could You Write This, Too...

There are many other product document types, some less popular than they once were. You may be asked to write any (or all!) of these:

- **Getting started guides.** Like a tutorial, a "Quick start" or "Getting started" document is designed to get a user up and running immediately. Many companies have discontinued these in favor of tutorials, and some include the "getting started" information inside their user guides.

- **Marketing communications (marcomm)**. This could include anything from press releases or data sheets, to the company's annual report. Marcomm calls on many of the same skills tech writers use every day but requires snappier language aimed at the paying customer.
- **White papers.** A white paper in today's business world typically mixes marketing and technology to help readers solve a problem. A white paper may simply describe a company's new product or solution, or it may provide detailed research and background information, but its ultimate goal is to help sell your company's product. A white paper should be written in a clear, easy-to-understand voice; it is meant to be understood by a wide range of readers.
- **Training materials**. Training materials, as I discussed in Chapter 7, "It's All About Audience,"are often slide decks written at a high level (because the details are filled in verbally by the instructor in front of the class). Sometimes technical writers are responsible for writing the training materials; most times, you'll be responsible for writing the source material that trainers use to build their slide decks.
- **Reference documentation**. Reference documentation can refer to anything that is informational rather than procedural. It is used like a dictionary or encyclopedia, and referred to when a user needs information about a command or topic. A reference guide could be anything from a list of error codes or messages, a list of Linux commands, a glossary, a definition of HTML tags, or a Tech Pubs style guide.

It's a good idea to become familiar with all these different document types; sooner or later, every tech writer's middle name becomes "Versatility." But every deliverable, no matter how far off your beaten path, must still incorporate the fundamentals: it must be correct, complete, usable, clear, and consistent, and it must be all of these things to its user.

Part 3. The Best Laid Plans

…do *not* go astray if you are well-prepared. (If you're like many tech writers, you'll know that the real quote from Robert Burns is "The best-laid schemes o' mice an' men Gang aft agley.")

Although you're probably eager to get started right away, planning and preparation is a big part of the tech-writing game. This section helps you plan your project, its schedule, and the formats you'll be producing. It also helps you learn about the technology you'll write about. That's the kind of on-the-job training you want.

Chapter 9

Process and Planning

You can write documentation without planning. But you won't do it as well or on as precise a schedule.

What's in this chapter

- Five steps to producing technical documentation
- Agile versus Waterfall—huh?
- Building a documentation plan
- The importance of milestones

So far, this book has provided a lot of information about what you will do, but nothing about how you will actually do it. You must be wondering, *When do I start actually writing?*

Not quite yet.

Before you start writing documentation, it's important to plan what you're going to write and when you'll deliver it. Process is important, and planning is important. As much as it may feel like a drain on your time to develop a process and a plan, and to follow an outline, doing so will save time and improve quality.

At the simplest level, the steps for creating documentation are the same in every tech writing project you take on. Certain activities are necessary whether you are taking over a project started by someone else, or starting your own project from scratch, and each takes time. You may spend more or less time on any of the steps, but you must spend *some* time on each of them.

Step by Step to Final Documentation

Every documentation project requires all of these steps. But it's not as easy as simply going through each step and checking it off. Documentation development is an iterative process, requiring you to return often to previous steps.

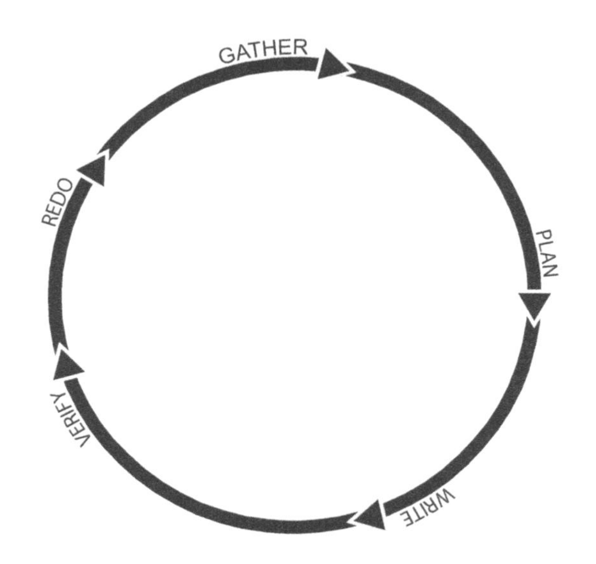

1. Gather information

Gathering information is ongoing. The entire time you work on a project, and afterward, you will gather information. It can mean everything from learning the product to interviewing developers to finding out what the customers complain about when they call for tech support. This can be the biggest part of your document development when you are new to a company—but it will be reduced as time goes on. Just remember, the more you know about the subject matter and the needs of the users, the faster you will be able to organize and create documentation.

Start gathering information as soon as you become aware of your assignment. Gathering information can take up to 70 percent of your documentation development time if you are new to a product; more like 10 percent once you become an expert.

- **Read** product literature, specifications, and engineering documentation to get a basic education about your subject matter.
- **Learn** by using the product so you can understand it thoroughly. Find out more about reading and learning in detail in Chapter 10, "Become Your Own Subject Matter Expert."

- **Understand** your audience by gaining knowledge of your users' needs. Chapter 7, "It's All About Audience" discusses how to do this.
- **Interview** subject matter experts (SMEs) and understand how the product works from the inside out. Chapter 14, "Gathering Information," contains pointers on interviewing.
- **Listen** while attending all the training sessions and meetings you can. Often the team forgets to invite the tech writers to the important meetings. Let it be your responsibility to find out where you need to be.
- **Be open** to receiving feedback and constructive criticism. Remember, your output is not a work of art; it's a work product designed to help the customer—whoever that customer is. If it doesn't succeed, keep your cool and make the changes you need to make.

2. Plan

Once you've learned enough to understand what you need to do to get started, it's time to plan and organize your information. You definitely don't want to wait until you know everything there is to know—remember that gathering information is never-ending, even while you're working on a schedule!

Planning can take up to a few days depending on the size of your project; however, a good document plan can carry you through several development cycles. Fortunately, subsequent releases will require much less planning than the first one does.

- **Determine** what your deliverables are, based on audience needs. Will you produce web content and online help? A printed quick-start guide and several PDF documents? Knowledge base content and a regularly updated Twitter feed? There are many types of documentation, as discussed in Chapter 8, "The Deliverables," and many ways to produce it, as discussed in Chapter 12, "You Want it *How?*"
- **Schedule** your work. Decide what your milestones will be and make sure you can meet them in time for the final delivery. Chapter 11, "You Want it *When?*" contains some tips for scheduling.
- **Post** a document plan similar to the one described "The Documentation Plan" on page 112. Even if it changes later, it's good to put a stake in the ground now. Letting your team know what you plan to do and when avoids mismatched expectations later.

3. Write

It can't be put off forever. At some point, you need to start writing! Time spent up front in learning and planning will greatly reduce the actual writing time.

As you become more of an expert at the product, familiar with the product development lifecycle of your company, and as the product goes through subsequent updates and releases, you'll have a higher percentage of time available for writing.

- **Outline** your work. Create a high-level outline of all topics that need to be covered. Discuss it with your stakeholders to make sure you've included the right information. Then create subtopics. Then create sub-subtopics. If you continue this way, you will eventually complete the entire project.
- **Start typing** even if you don't quite know where some information will go. Don't worry if your wording isn't perfect; the important thing is to get some words down. Chapter 15, "Putting It All Together," helps you with that.
- **Create placeholders.** If you don't yet know what exactly will go into a specific section, just enter a heading that states what the subject will be, and a line that reads "TBS" (To Be Supplied) or "Will fill this in later." Placeholders are very useful to you and your reviewers, since it means that you know that these subjects need to be covered and the important information will not be overlooked.

4. Verify

Verification refers to the stage where your work is tested. This means both by you and your subject matter experts.

Verifying your documentation will take different amounts of time depending on how much verification you must do (testing an installation by doing it yourself, for example) and how much verification is required by reviewers. The complexity of the subject matter is also a factor here; obviously, a highly technical subject will require much more reviewing.

- **Test** your documentation by reviewing it yourself. Go through all the procedures and make sure they work. If you are writing installation instructions, perform the installation yourself. If you are documenting a user interface, match your documentation against the UI. If you are creating help topics, test links to other topics.
- **Proofread** your work to make sure the language flows and content is complete.

- **Review** your documentation by distributing it to subject matter experts and asking them to read it and confirm its accuracy and completeness. Chapter 16, "Everybody's a Critic—Reviews and Reviewers," talks about the review process.

5. Redo

With every review, every reading, and every customer call, you will find something that needs to be revised, cleaned up, or improved in some way.

- **Correct** your documentation by fixing errors and typos.
- **Clarify** your documentation by making it easy for the user to understand and find information.
- **Rewrite** your documentation to improve it and add missing information.
- **Retest** the content by returning to step 4, Verify. You may need to send it out for another review, or you may need to read and check your own content. Continue this sequence as many times as it takes to get it right, or until you run out of time.

And, of course, as subsequent releases come out, a great deal of your documentation development time will be spent redoing existing documentation and updating it with new information.

Understand the Development Process

The documentation process is important, but you also need to understand and follow the *development process* your company uses when it develops or manufactures software or hardware. If you work with software developers, make sure you are familiar with and part of your organization's software development lifecycle (SDLC) process. Writers frequently work on multiple projects across more than one product and may have to follow more than one development process.

The software development process can have a huge effect on your working life, as you will see from comparing just two development methodologies: Agile and Waterfall. An information developer (you) should work with the same methodologies as your development teams. That means you should also know what they're all about.

The Agile Process

Agile software development is a methodology based on iterative development, teamwork, and open communication. Agile development teams frequently follow the *scrum* process, in which development teams work in short *sprints*, typically two to three weeks, during which they are expected to focus on and fully finish a set of features. Features and functionality are defined in a prioritized list of requirements called a backlog. At the end of the sprint, the software is considered to be ready to deliver, so at any time, the team is never more than three weeks from the ability to deliver a product that could theoretically be delivered.

Each sprint is preceded by a planning meeting—where the tasks for the sprint are identified and the team members make commitments—and followed by a review, during which the team's progress is demonstrated and the features are shown to work.

The tech writer as part of the team

Agile sounds great for the tech writer, right? And it can be, when the writer becomes a key part of a dedicated scrum team, documenting each feature as it's being developed, and working in an environment in which a task is not considered complete until the documentation is reviewed and finished. In an ideal Agile world, by the time the writer is ready to generate help or assemble a manual, all of the pieces have been determined to be accurate and it's just a matter of "some assembly required."

But there are also pitfalls for a writer working in this environment. The writer is often also assigned to many other projects that don't necessarily align with the scrum teams. In a working environment where team members are often expected to be dedicated, it is difficult, if not impossible, for a writer to find the time to be part of all the scrum teams she should be on. When the developers on a scrum team are dedicated to the current or upcoming release, they aren't necessarily aware of the fact that the writer also must maintain and correct old documentation for releases long gone and for projects being developed by other teams.

Question: *In a bacon-and-egg breakfast, what's the difference between the chicken and the pig?*

Answer: *The chicken is involved, but the pig is committed.*

And that's meant to sum up the members of an Agile scrum team. Pigs are totally committed to the project and accountable for its outcome, and chickens consult and are informed of its progress. It may be the one time in your life when you'll be glad to be called a pig.

A writer must make certain that all documentation tasks are accounted for in the sprints—not just writing, but also ensuring that the team members review the content. It's not easy for the developers to review at the same time they are developing a feature, and there is a lot of time pressure for the writer to document the feature in time for the review to occur during the scrum.

Because of that, the documentation does not always get reviewed on time. Writers are traditionally accustomed to waiting until features and the user interface are complete before producing documentation. This does not work in a scrum environment and the writer must rethink the way she's always done things.

As well, documenting a feature while it's being developed is not necessarily the right thing for a technical writer to spend her time on. Sometimes a development team is working hard on a set of features and the scrum team expects them to be fully documented, yet they are internal functions that will not even be visible to a customer. Or the large development effort, and hours of the writer's time as she follows the scrum team, translates to a single sentence in the user guide.

All of these issues are valid, and must be worked through with your team so you can all work together to produce good product documentation. You may decide to create *user stories*—short statements about what the user does with the product—that are specific to documentation, such as:

> The user can get help on the topic by clicking a Help link next to the text field.

If you are writing documentation that requires the user interface be finalized, you may decide that documentation tasks always follow one sprint behind the development tasks. Or you may decide that you can do the documentation during the sprint but the review will wait until the documentation components are assembled at a later stage. Work with your team to determine the best way to achieve this.

Final documentation work is also a sprint

Scrum teams sometimes forget about documentation as they finish the work for the sprint: there is almost always a set of document production tasks near the end of the development cycle. At some point, you must assemble the bits and pieces you've written during the sprints and turn them into a deliverable—a help library, a guide, web content, or all of the above. The team must respect that these tasks must occur and you must make sure that you plan for them.

All in all, Agile (and scrum) can be richly rewarding for a tech writer, but you must be aware of the potential problems and work with your team to resolve them. Luckily, if things aren't going right in this sprint, there will always be another upcoming sprint to try to resolve issues that are not working for you.

The Waterfall Process

The Waterfall methodology is the more traditional method of software development, in which design is done sequentially, through phases of concept, initiation, analysis, design, construction, testing, implementation, and maintenance. All work must be completed in each phase before another phase can begin.

Writers accustomed to working with the Waterfall model will typically determine at which point they need to start getting involved, and they will work on their own documentation time line, planning to deliver their final product at the same time as the product is released. This system allows the writer to work independently.

While this method works well for many tech writers, it also has its down side. The writer is not really part of the team and sometimes acts more like a visiting reporter. The writer frequently doesn't learn how the software behaves until very late in the cycle and then has to rush to put all the information together at the end. Developers can resent having to explain and re-explain information that they have been immersed in while it seems as if the technical writer isn't really involved.

If your organization uses both Agile and Waterfall methodology, it can be difficult for you, the writer caught in the middle. You may find yourself working on an upcoming product with the Agile group and the product that's already finished for the Waterfall group. And neither of those two development teams will be interested in the fact that you're working on the other's deliverables.

The Documentation Plan

A documentation plan is a useful tool, not only for helping you plan your work, but for letting your stakeholders know just what you intend to do. The documentation plan sets expectations for you and the whole team and gives you a structure to work from. As you build the plan, you start to visualize how much work is involved in a project and how much time it truly takes.

A documentation plan may be as simple as an email that states your intentions in a few sentences, or it may be a detailed document. In any case, it should spell out how you intend to get from where you are now to where you need to be. It can describe a single documentation deliverable, or it can describe a set of documentation that supports an entire product family.

If you've never created a documentation plan before, start with the basics:

- **Name of the deliverable.** This may be anything from a specific document title like *EnterPrize Cloud Storage Version 2.0 Installation Guide to* a more general description like "Web content for buy pages."
- **What needs to be done.** Include a description of what is required for this documentation: whether it's to be done from scratch, an update for a new release, or it needs to be generated as several different types of content formats for different operating systems or devices.
- **Writer working on the deliverable.** If you are not the only writer working on the project, create a column to indicate who the writer is for each deliverable. If it's just you, make that clear at the top of your plan.

- **Reviewers.** List all reviewer names. Make sure you get buy-in from them.
- **Milestones and their delivery dates.** If you aren't sure exactly what they are, make a guess. It gives you a starting point for discussions with product owners and developers.
- **Comments**. Leave a space for comments, even if you don't have any now.

A documentation plan doesn't need to include a lot of detail, as the following very basic example shows:

HomePrize Cloud Storage 2.0 Documentation Plan			
Deliverable	*Reviewer*	*Milestones*	*Comments*
User Guide New content for release. Additional appendix for keyboard shortcuts	Rob, Andrea, Lin	Draft 1: June 3 Draft 2: July 10 Final: July 16	Waiting to see if the Administrator function is being added.
Help Single-source from user guide for output on website, desktop software, and mobile device	Rob, Andrea, Lin	Final: July 13	
Installation Guide New section on upgrading from 1.0, appendix on user permission	Raj, Josh	Draft 1: June 21 Final: July 6	
Release Notes All new	Rob, Lin, Raj	Final: July 18	

Some writers like to create very detailed documentation plans, a different one for each deliverable, especially if they are producing new content from scratch or writing for a new product. They use the documentation plan to define the purpose, the content, the target customers, the priorities, and more of all the documentation. A detailed documentation plan can be reviewed with all the stakeholders to make sure everyone has a common goal.

If you are working in an Agile environment, you may think that a documentation plan won't apply to your situation. Not so. The documentation plan is a great way to see the big picture—all of the documentation that needs to be done—and then figure out where your tasks fit within the sprints. Make sure your plan meshes with the sprint planning.

Milestones

The milestones you need to identify for each project (and put in your documentation plan) are:

- Draft(s) distributed
- Reviewer comments due back
- Final handoff

There may be several rounds of these; if so, put them all in the plan. For example, you may be planning a brand-new document for a complicated product and realize that one draft won't be enough for something like that. Or perhaps you'll want to send out a draft for a single chapter or a single topic. You'll count on having to make changes and sending out another draft for review. Put them all in the plan.

Originally, milestones were actual stones set into the ground beside primitive roads. They named the nearest city or town and the distance to it. Milestones in tech writing perform a similar function. They are the interim goals you must reach before you go the distance.

The documentation plan helps you get buy-in from project stakeholders. Make sure that after you put the plan together you send it out for approval to the people whose participation you need: people like your manager, the product manager, and the engineers on your project.

The documentation plan has many advantages, some of which I hope you won't need. If it happens that someone claims you didn't do something you were supposed to do, that the content was wrong, or that you missed a date, there is nothing better than pointing to a publicly posted documentation plan that you were fully above board in reviewing with the team. It certainly doesn't hurt to spend that bit of extra time up front working on something that not only helps you work better, but can also help cover your tracks in a case of emergency.

Chapter 10

Become Your Own Subject Matter Expert

Mastering the ins and outs of your company's product—it's what you need to know to be able to really write technical documentation.

What's in this chapter

- Why you need to learn the technology
- Keeping the "technical" in technical writer
- The one—and only—time there is value in ignorance
- Letting yourself learn

When you begin a job or are assigned a new product to document, it often seems that many things are preventing you from meeting your deadlines. The obstacles might be other people who aren't holding up their end—a hardware prototype is running late, or you weren't given the appropriate equipment or user account to run the software you're writing about. Or the obstacles might be your own fears—you have no understanding of this technology and yet you're expected to document it for people who know more about it than you do.

If you're new on the job, you don't even know yet how to get around the office building, and the hundred or so acronyms you've heard are swirling around in your head like alphabet soup, how can you expect to learn the ins and outs of a complicated technology in time to meet a crazy delivery date? This can be a challenging situation. But don't let it stop you from jumping in to learn as much as you can, as soon as you can. This chapter will help you learn how you can come up to speed on the subjects you're expected to know.

The Importance of Product Knowledge

To produce fast and accurate work, you must have a substantial amount of product knowledge. That isn't to say that your co-workers will expect you to know everything as soon as you walk into a new job. It would be nice if you had enough experience with the given technology to know what questions to ask, but even that isn't always going to happen. When you start work, be honest about what you don't know. Ask for help where you need it, and learn what you can by using some of the techniques described in this chapter.

Instead of depending on subject matter experts for everything, you can *become* the SME. There are many reasons it's a good idea for a technical writer to gain extensive product and technical knowledge:

- You will gain the respect of the developers and be able to talk to them with some intelligence. (This is a biggie.)
- You will be able to determine whether information you receive or read is correct. (This is also a biggie.)
- You will be able to write much faster than you would if you know nothing.
- You won't ask the same questions over and over. (Which developers hate.)
- You will have confidence that your documentation is accurate, even when it has not been sufficiently reviewed by subject matter experts.

When you become well-versed in the technical and user aspects of the product about which you are writing, you will find it much easier to write. Compare this to a writer who is tasked with working on a software installation guide but has never installed the software and has no way to install the software. This writer has to run to Engineering or Quality Assurance (QA) or Operations and ask them what happens at various steps. He must ask someone to go through an installation while he watches and attempts to take notes.

That kind of documentation requires numerous reviews and rewrites. And it's bound to be wrong because the reviewers are unlikely to test every step in every procedure against a real installation to fill in all the holes for him. If they are expected to do all that writing and testing, why does the company need the technical writer? Just saying…

You might feel, resentfully, as if your boss expects you have the knowledge of an engineer when you're "only" a technical writer. That's not likely to be true, but you should have some expertise in every area of the product you are documenting. As the assigned writer for a product, you should strive to have the knowledge of:

- **An expert user** when it comes to using the product
- **An intermediate user** regarding the purpose of the product. (Writing about image-editing software? Learn what retouchers do and expect to do. Writing about medical equipment? Understand what its operators need to know.)

> INSIDERS KNOW
>
> It's tempting to just nod and pretend you get it when you hear something you don't understand—but don't do it. You may think you look smarter, but it can have disastrous consequences later. Busy people hate having to answer the same questions over and over.
>
> When you encounter something you don't understand, admit it and ask the subject matter expert to explain it to you.

No one's asking you to gain all the knowledge and expertise of every professional in the company, but you should also strive to have at least a junior-level, or "dummies"-level knowledge of each of the following:

- **A Product Management** understanding of how the product behaves, looks, meets requirements, and serves the needs of the customers
- **A Quality Assurance (QA)** understanding when it comes to being aware if something doesn't work as expected
- **A User Experience** understanding of who the product user is and how easy the product is to use
- **An Engineering understanding of** the technology behind the product
- **A Business Development specialist** understanding of how important this product is to the business plans of your company. Are there a lot of customers? Are they important customers? Will this product have a future release? Are there plans to expand the product?

Put the "Technical" in Technical Writing

There was a time, thankfully short-lived, when a school of thought held that tech writers shouldn't know anything about the technology or product that they wrote about; in other words, they should go in fresh, just as an ignorant user would. These untainted tech writers would then proceed to churn through the menus and buttons on an application and document everything they saw, exactly as they saw it. Or they would go to the developers and ask them to write about the product and then the tech writer would edit and format the content.

There are many problems with this way of thinking. The main problem is that documentation written by a writer who knows nothing offers nothing the users can't get for themselves—less, actually, since users have personal or business reasons for acquiring this product, and therefore already have some sense of what they want to do with it.

> **INSIDERS KNOW**
>
> If you feel a need to change a technical term to something more "user-friendly," it may mean that you don't understand its use. It's happened more than once that a well-meaning technical writer has changed punctuation marks in UNIX code because the writer didn't know they were integral parts of the command. Changing a word that sounds odd to you can make an instruction or code sequence meaningless.
>
> When you find a term you don't understand, do a web search on it first before you ask someone for more information.

At a minimum, what the technical writer should be able to provide is an insider's knowledge—that is, an understanding of how to do things that are not immediately obvious. That means knowledge and guidance about what is to be entered into various text fields, and an explanation of what settings are defaults and what defaults should be changed.

When a tech writer documents her own first experience, unless she is continually asking and getting answers to the important questions—What's the maximum number of characters I can put in this field? What's the difference between the Submit and Apply functions on this screen? Why would I need to ping the server at this point? What kinds of cables can I use with this?—she is providing nothing that the end users couldn't see for themselves if they too plodded through each menu and screen.

It's OK to write for a naive user. It's not OK to *be* a naive writer, so you'll want to get out of that mode as soon as possible.

Read Existing Documentation

Remember all the documents described in Chapter 8, "The Deliverables"? If product documentation exists for this product already, read it. Start with the marketing and sales material, pretending you're a potential customer, and look at the high-level documents like product sheets, data sheets, and other material that your sales or marketing department distributes.

If the product is still in the development stage, ask for requirements, design documents, and CAD drawings.

If you're like most of us, just reading the words won't be enough. You'll need to actually see it for yourself.

Get a Demo

Before you try using the product yourself, ask someone who understands it really well to give you a demonstration. This is likely to be the product manager, who is usually the owner of the product, a QA person, or developer who works on it.

Because you have read the documentation, you should have a sense of the purpose of the product as well as an idea of how it works. Be prepared with some questions: What are the users' goals? How does this solve their needs? Are there different ways to do the same thing (such as keyboard shortcuts)? How does the user know what types of information to type into the text fields?

Do it Yourself

If the product you are writing about is still in the development stage, you may have to settle for reading about the product or seeing pictures of it rather than actually using or touching it. But once the product is available, you want to make sure you have what you need to succeed. If you're writing about software, this means:

- **A "sandbox" or login environment where you can install the software yourself.** Make sure you can access the system when you need to without interfering with someone else's work or having them interfere with yours. This might mean Tech Pubs needs its own environment (or you need your own) instead of sharing with Development or QA or another part of the organization.
- **The same user privileges your customer will have.** Ask for administrative privileges also, to allow you to manage users.
- **Dummy directory structures and data.** This lets you set up examples and screen shots that resemble those the customer will make or see.
- **Access to the latest and greatest version or release.** A demo version won't do. You need access to the real thing, even if—or especially if—it changes every day.

If you're writing about hardware, you will want a device of your own to see and touch and play with. If you're writing an installation document or wizard, you'll need to install the product yourself.

Play Around

Sit down with your product, something to take notes with, and a cup of coffee. You're ready to dig in. Now what? Whether it's software, hardware, or a combination of both, start playing around with it to see how it works. There are a couple of approaches you can take.

You could take a methodical approach, which begins at the first screen and works sequentially through every menu, button, and page if it's software you're looking at. (To use the methodical approach on a hardware device, push every button and switch and explore every part.) The benefit of this approach is that you are likely to see everything there is to see. You are likely to discover functionality that you did not know existed, or more than one way to perform a function. The drawback of this method is that you may not understand the user tasks.

A common complaint of developers is that technical writers keep coming back and asking the same questions. Don't be one of those writers—you wouldn't like it if someone did that to you, would you?

Before you even go to a developer, make sure you understand as much as you can, and have specific questions ready to ask. When he tells you something, write it down and check to make sure you understand. If you need more clarification, go back for more explanation, but be clear about what part you don't understand so at least you're acknowledging what he did answer already.

A task-based approach

A task-based approach is one in which you first understand (or guess) what the user wants to do with this product, and then try to figure out how to use the product to accomplish the tasks. You remember reading about task-based documentation in "Task-Based: What the User Does" on page 90. Now you'll understand the value of having that type of documentation as you try to understand the user's tasks and workflows.

The advantage of this approach is that you're learning this product as the user would, but because you have access to the developers and the design documentation, you are able to ask all the questions the users might ask if they were there. The drawback to this approach is that if you don't fully understand what the user needs to do, you may not be able to put the product through its paces.

Understand the workflow

It's best to combine both of these approaches when you're learning a product. Learn about the workflows, and as you go through each component of a product, think about how they are used in the workflow. Solve user problems by looking for the best and shortest ways to accomplish a task. Go back to your experts and ask them questions.

As you ask questions and learn more, you may be surprised to discover that the developers in your company don't really understand how the customers will use the product. In many cases, software is built solely to meet the requirements. The developers fit as many features and functionality into the products as they think they can, and assume that an expert user will know exactly what to do with them. This is not particularly helpful in figuring out how to explain tasks to your users.

You're lucky to be a technical writer today, where the Internet makes fast learning so much easier. When you hear a word or acronym you don't know, look it up on **techterms.com** or **acronymfinder.com**.

Make sure you talk to someone (try Marketing or Sales if you aren't getting that information from Product Management or Product Development) who really understands the customers and what they need to do. If there's no one like that at your company, the next step is to try to learn what the customer knows.

Become an Expert in the Domain

In the high-tech world, *domain expertise* often refers to expert knowledge in an aspect of the business. You should strive to become a domain expert in the field you're writing about.

If you're writing documentation for network administrators, learn about networking. If you're writing about Internet security, learn about digital certificates and how they work. If you're documenting financial software, learn about the tasks your user expects to do.This can be difficult—more difficult than learning the technology that lies behind your product—and it will take time, but even a small amount of this knowledge will be extremely helpful in producing task-based documentation.

Check out the Competition

It's always a good idea to keep up on competing products. Read the documentation written for products that are similar to the one you are documenting, and learn from them. Don't steal any content or copyright material, but you can certainly pay attention to the way the documentation organizes its material and describes tasks, and learn from the content.

Capture Your "Newbie" Experience

Remember, as you first use the program or device, that this is your first and last chance to experience it from the perspective of the naive user. Take advantage of that. Be aware of what you don't know and how you search for information, and make notes of the things that are difficult or non-intuitive so you can plan to describe them in some detail. In fact, you should write everything down—you'll need it later when the product becomes familiar to you and you forget how it was to use it as a novice.

This is the one time when even ignorance can be useful.

It Takes Time

Give yourself time to learn the product, and don't get discouraged if it seems to take too long. Everyone at the company, no matter what their job, took some time to learn what they know now. As time passes, you will know more, and by the second release of your documentation, you will already start to feel more like the expert you'll soon be.

Chapter 11

You Want It *When?*

How to keep up in a notoriously fast-paced industry.

What's in this chapter

- When good, fast, or cheap becomes just "fast"
- Redefining "good"
- Learning how to work with contractors
- Working backward to plan a schedule

Fast, Good, or Cheap: Pick Two

You've probably heard this saying or seen it on a shop sign: "You can have it fast, good, or cheap: pick two." Like a lot of old sayings, there is much truth behind it:

- You can provide high-quality work with quick turnaround if you spend the money to hire more people—fast and good, but not cheap.

  Or:

- You can write excellent documentation that meets the customers' needs, and do it all by yourself, but it will take so long to do it, the next product version will be on the shelves before you finish—good and cheap, but not fast.

  Or:

- You can do a project on a crazy schedule and do it alone, but the quality will not be the best—fast and cheap, but not so good.

But times do have a way of changing. In the high-tech world, companies and customers want their documentation to be good...and fast...and cheap. Is it possible? Especially when everyone has their own ideas of what these all mean.

Yes, it is possible, if you amend the saying, to, "You can have it good enough, fast enough, and cheap enough. Pick the most important one and let's readjust your perceptions on what the others look like."

Good Enough Has to Be Good Enough

In many companies, "fast" will be the most important, and "cheap"—that is, using the fewest number of paid people to produce the largest amount of output—is often a given rather than a negotiable in today's world.

So that leaves us with "good enough" instead of "good." But what does that mean in the high-tech world?

Writing high-quality documentation is always important, isn't it? Or is it? Many people, if asked, would say that of course quality is the most important thing in documentation. But to gauge its real importance, consider a scenario in which the documentation team, for whatever reason, can't complete documentation in time to meet a tight deadline.

New tech writers often don't understand that some types of documentation are more important than others. A writer might mistakenly expend the same amount of time and effort on a manual that goes to a limited set of customers as she will on a manual that documents the company's biggest-selling product. Make sure you understand your business and its priorities. When in doubt, ask your manager.

Does the manager push back the deadline and tell the writers to continue working until the content is complete? Or are the writers instead instructed to do as much as they can by the deadline and put out the documentation in whatever form it's in, waiting until later to add missing content? It's quality—or the perception of quality—that is subject to adjustment.

In an industry in which time feels measured in microseconds—or in nanoseconds, at Internet-driven dot-com companies—speed is often going to be the most important of the three criteria. Development schedules speed up and companies produce more products and try hard to please many different customers. The days are gone when a tech writer could spend six months working on a single short document. Today it is not unusual for a writer to be responsible for one or more product families and all the associated documentation, and the writer is often given a matter of weeks—or less—to deliver a finished product.

Because tech writers care about their work and want to produce the best documentation possible for the customer, they often find themselves frustrated as they work on tight deadlines with overloaded schedules to produce the highest quality documentation possible within the given time frame. That means working extra fast, working longer hours, and working at a higher stress level as they strive to wrap everything up.

Battling Your Inner Perfectionist

When speed and price are more important than quality, expect to be frustrated. This is because it seems to be part of the tech writer's nature to want things to be perfect. And that's not surprising; after all, we're detail-oriented and service-oriented. We like to explain things to people and we are interested in completeness and accuracy. We care about the placement of commas, when to say "login" versus "log in," and can argue forever about the way bullet lists are formatted.

But there will be times when you'll have to learn to let go of your idealized vision of perfectionism, without giving up on quality completely. Think about quality as consisting of five factors, listed here in order of importance:

1. **Accurate.** Every single piece of information is true and described unambiguously.
2. **Complete.** You have anticipated what the user needs to know and included it all.
3. **Error-free.** No typos, no bad formatting. Headings and terminology are used consistently.
4. **Usable.** Well-organized and presented in a format that suits the product and its users. (We'll learn more about usability in Chapter 12, "You Want it *How?*")
5. **Good out-of-the-box experience.** You don't have as much control over this, and it's definitely more important to companies that favor "good" over cheap, but your documentation can help.

Your documentation can live without numbers 2 through 5. But it absolutely cannot do without correct—that is, accurate and error-free—content. So if there is nothing else you focus on, make sure you focus on *accuracy*. In the long run, it will save you and the company the most time. You, because you will have fewer complaints from internal and external users and fewer revisions to make. The company, because there will be fewer calls to customer support.

Get Your Speed On

Some people work faster than others. Prolific fiction writers like Michael Connelly or Ruth Rendell knock out two best sellers a year, while some of us stare at our computer keyboards and can't seem to get started. Whether you are a tortoise or a hare, there are a few tricks you can use to work faster:

- **Start sooner than you think you need to and do a little bit each day.** Almost every writing project takes longer than you expect it to, so it's important to avoid a last-minute crunch by starting as early as you can and working steadily on the project. If by some miracle it doesn't take longer, think how nice it will feel to finish ahead of schedule.

- **Break down your schedule.** Break down your schedule into the smallest chunks you can. If you are working in an Agile environment (see Chapter 9, "Process and Planning") you already have a good starting point, since your deliverable depends on the work being done for that sprint. But you can still make milestones for yourself within the sprint. You may want to create daily milestones for yourself and make sure you meet each of them each day. This also helps you work at an even pace—like a marathoner, you won't burn out early and you'll save some energy to be able to rally at the end.

- **Create an emergency.** I'm not kidding. If you're one of those people who waited until the night before a college term paper was due to start writing it, you may be tempted to do the same with your tech writing assignments and then be caught short when circumstances beyond your control create havoc.

  Set a deadline for yourself and make sure others know about it. Before you finish the document, send email to the reviewers telling them you'll be sending out a draft on a certain date and you will appreciate their attention and speedy turnaround. You might even itemize some of the topics you'll be covering in the draft to force yourself to fill in the gaps you haven't gotten to yet.

  Be careful, though. If you don't come through on what you've committed to, colleagues will think of you as someone who is not reliable and reviewers won't feel there's any particular responsibility to respond on time.

- **Make a list.** Some of us are list-makers and some of us aren't, but for those of us who are, there is nothing more satisfying than crossing off the finished items. Start your day or week by making a list of all the tasks you must complete, no matter how large or small. Enter it into your email program's task list or write it on a large sticky note—whatever style works best for you. Review your list regularly and add to it as things come up. This is helpful in ensuring you complete your work and it also gives you a much better sense of the total amount of work you've got to do and the time in which it needs to be completed. If the list gets out of hand, discuss it with your manager.

Multitasking

Did you ever see a circus juggler who sets plates spinning on top of poles and then rushes madly from pole to pole to keep them all spinning? This is the way a multitasker sometimes works, and succeeding at it may be the true secret of people who work at lightning speed.

When a "unitasker" has multiple projects going on, he works on one project, or one activity, at a time. This writer will work on a single task until he feels it is complete before picking up the next task.

A multitasker has a lot of activities going on at the same time. If tired or blocked on one task or project, the multitasker switches gears. Without a pause, the multitasker jumps into a task that is significantly different from the original task, such as indexing, proofreading, or working on a completely different project.

Working at breakneck speed means doing something productive all the time, and often appearing to do several of them at once: these people proofread hard copy while waiting for the computer to start up, index while generating help, and edit while talking on the phone. In the same time it takes to go refill your coffee cup or ask about a co-worker's weekend, the multitasker has crossed several items off her to-do list.

We've probably all heard the research that tells us that humans can't really multitask. And all we need to do to prove it is to go to any meeting where half the attendees are web-surfing or texting and ask them what was just said. What successful multitaskers really do is rapidly switch attention from task to task.

Refer to your list to have other tasks ready to do on the spot, the moment you need a break from your current one, or you hit a stopping point. It takes practice, but you can significantly increase your productivity.

Turn Off the Email

It's hard to focus on a single task when popups are constantly clamoring for your attention. Acting as a human auto-responder doesn't count as productive multi-tasking; it can be a distraction that makes you work more slowly.

In his book, *The 4-Hour WorkWeek*, Timothy Ferriss tells us that he only reads his business email for one hour each Monday. You don't need to go to that extreme, and your manager is bound to be annoyed if you do, but you can definitely benefit from Ferriss's excellent suggestions for limiting email, which he calls the "greatest single interruption in the modern world." If you find yourself auto-responding to your email alert even when you have far more important things to work on, turn it off. Check email twice per day, once before lunch and once around 4:00 PM. Ferriss suggests that you never check email first thing in the morning, and instead complete your most important task before 11:00 AM.

If you decide to reduce your time spent reading email, make sure your manager knows about your routine and has a way to reach you if necessary.

Working with Contractors

When tech writers have too much to do and their budgets allow, it's a common practice to hire contractors, or short-term hourly workers, to help out. As the saying goes, many hands make light work.

Of course, there's another saying, also true: "Nine women can't make a baby in one month," which reminds us that there is only so much help that a contractor can provide. This section discusses some of the things you need to know, both positive and negative, when bringing temporary workers into a writing project.

Contractors can help if you are able to carve out a significant amount of "extra" work; that is, work that does not require product knowledge but still needs to be done. If you have enough indexing, proofreading, template-building, help-generation, and other tasks of that nature, give them to a contractor and let the regular staff work on important core work.

People who can handle that type of work are also usually less expensive than the ones who have the domain expertise, and as they become familiar with your products, they will eventually be able to handle the harder projects.

There are many tech writers who make their living working as contractors. They are typically senior writers who have expertise in certain subject matter and are well-versed in the tools used by tech writers in your area. They are comfortable stepping into a new situation, and are able to work independently. And since they are not full-time employees with benefits, the money to pay for them usually comes out of a different budget than the one used to pay employees, a budget with more flexibility.

Paying the Price

Hiring contractors during an emergency situation can be an unexpectedly expensive way to get the job done, and is sometimes more time-consuming than time-efficient. Although a good contractor should be able to step in and hit the ground running, the fact remains that she still needs time to come up to speed. And you'll find that the extra time required of you, or your manager, and your co-workers to train the newcomer can hurt.

You may think that hiring a person who understands your technology or has worked for one of your competitors will eliminate the ramp-up time. That won't happen, although the time is likely to be reduced. But finding the skill sets you are looking for can be costly. Technical writers with experience in networking, or mobile devices, or server management, or other specialized areas charge a high hourly rate.

If you're under pressure, the deadline is looming, and you're thinking you need to hire some temporary help to enable you to make your dates, step back and think it through. Are contractors really your best bet for meeting the upcoming deadline? Can you afford to spend a considerable percentage of your time acquainting them with your building and the people they need to work with, showing them the product, explaining your department's style and procedures, and making sure they understand what needs to be done and how?

Once the contractors are productive (estimate a week before they are able to help with tasks like indexing and proofreading against your style guide; a month before they are able to start adding valuable content to more technical documentation), is there still time to deliver? If you can answer yes to these questions, be glad that your company does not consider "cheap" to be the most important factor in document production, and go ahead and hire a contractor.

Anticipate Your Needs

The best way to avoid being trapped into a situation where you need a contractor but can't afford the time it takes to train one is to plan ahead. This means keeping close tabs on how a project is progressing and what is coming up in the immediate future. Doing so lets you anticipate your needs well in advance so you can make a case for contracting help in enough time to factor in the ramp-up time. If it takes a month to get someone on board and a week before they are even useful, that's fine when the deadline is eight weeks away, but not when it's two or three weeks away.

Make a case with your management as soon as you know that an upcoming project will require more resources than your department currently has. Emphasize that you need to get the contractor on board before the project starts.

Keep the following points in mind when interviewing or writing a job description for temporary help:

- **Make sure the person you hire is familiar with the tools you use.** It makes no sense for you to pay someone to learn Adobe FrameMaker or Robohelp.
- **Seek a contractor who is familiar with your technology.** If you want someone to work on run books for your NOC (Network Operations Center), make sure the person has data center experience. If your department writes

wiki content or web content that is heavy in JavaScript, make sure the candidate knows the technology and tools that will get the work done.

When planning to hire full-time or temporary help for an upcoming heavy workload, be aware that it will probably be five weeks or more from the time you open the requisition until you have someone sitting in a chair. Interviewing candidates takes a lot of time and if you hesitate, you can lose a good one, as desirable candidates often have their pick of offers.

Avoid the interview time-trap by networking in advance to build a pipeline of potential contractor help that you can call on when you need them, or build a relationship with an agency or agencies that specialize in technical communicators.

- **Make sure you and the contractor know each other's expectations.** A highly paid contractor should be equal to any technical communications task, whether it's writing from scratch, meeting tight deadlines, or dealing with engineers to dig out information. However, what you think the contractor will do and what the contractor thinks he'll do are not always the same. Before you sign the agreement, find out what the candidate will and won't do and make sure you are both in agreement about the expectations of the job. Discuss your expectations about where the contractor will work—whether it is on site or whether you are OK with them telecommuting.
- You don't want the contractor to walk out on you the second day of the job because you ask them to do something they don't think is part of the job.

These guidelines do not always apply when you are hiring full-time help. Full-time employees are an investment and for the right person, you can expect to invest time in training.

When you find the right candidate and have a start date lined up, congratulations! With the right expectations in place, you both have a good chance of having a productive and long-lasting working relationship.

Short-Term Productivity

Once you have the contractor in place, follow these guidelines to ensure productivity:

- **Remember the contractor needs time to come up to speed.** If it's a short-term project and the deadline is tight, don't expect the contractor to step in and work magic by writing detailed technical documentation on the spot.
- **Make sure you have work ready to do.** Often when contractors come on board, the regular employees are so swamped, they have no time to deal

with the contractors. They put the newcomer in a cubicle with little to do—out of sight, out of mind. Prepare specific tasks in advance that make it easy for your contractor to get started.

This also lets you see how the newcomer is doing. Ignoring someone for a while during a busy period often means that you take much too long to realize when someone is not working out.

- **Keep your expectations realistic.** The contractor is not you, and does not know the product as well as you do. The contractor also has less at stake—to her, this is just another project. For you, your job performance may be on the line.

If you find someone who works well with your team and does a good job on your projects, do what you can to keep that person around. When you build a good relationship with a contractor, she will often give you "first pick" for jobs, and what a relief it will be to skip the lengthy ramp-up period next time you need a temporary worker. The demand (and price) for a good contractor is high. Providing steady work is a sure way to keep a contractor's attention.

Making and Following a Schedule

In the high-tech world, everything has to be done yesterday, if not last week. No matter how fast you start, you're already behind, and it seems there's never time to do the job right. But doing a job "right" can sometimes depend on recognizing what "right" looks like.

You'll never have unlimited time to produce documentation, and as discussed, you need to be able to keep up with the product development lifecycles your company follows. Typically, you will be expected to meet a deadline someone else has set and you'll have to decide what you can deliver in that time.

You don't have to react like a deer in headlights when asked if you can meet deadlines. Project planning has only three dimensions: size, scope, and quality. Planning a project is a matter of understanding how these dimensions interact.

Project Scope: How Wide Is Big? How High Is Up?

Whether you're planning an entire documentation set or a single deliverable, you need to understand how to schedule. Unfortunately, the scheduling "method" many people use is to nod their heads when told of the deadline, and then cross their fingers and continue at their standard pace, hoping it all works out at the end. While it's nice to have such confidence in your abilities, this is rarely enough.

The most important aspect of figuring your schedule is to understand the scope of the project. Luckily, scope can always be changed. If you have more time than expected, you can do more. If, like most of us in the high-tech world, you have less time, you can reduce the scope, either by producing less content, a less-complicated delivery method, or fewer components of the project.

Prioritize Your Deliverables

Review the items you believe are necessary and list them in order of priority. In other words, what items are absolutely essential and what are nice to have? Let's create a sample list of deliverables in what we believe is the order of priority for our fictitious company's new consumer product release, HomePrize Cloud Storage 1.0:

1. Web page content
2. How-to help
3. Troubleshooting guide
4. Quick-start guide
5. User guide
6. Developer's guide

Consider the Medium

We're not talking about psychic abilities here, although when planning a schedule, you might feel it would be helpful to have a psychic on hand. We're talking about the medium that conveys your content to the user.

Once upon a time, it was a given that documentation meant that you would almost surely produce a printed book, perhaps in a binder, and you had to allow a lot of time for that. Today production applies to everything from PDF files to generating help to putting your content into a software build to creating a multimedia piece to pushing to a web server. Before you plan your milestones, think about how much time that end result will take and factor it in. If it is a big piece of your available time, consider producing something simpler.

Review your list of deliverables and estimate the amount of content and the scope of effort needed to produce each item. To determine scope, ask yourself questions like the following:

- For the website, will you need to write original content, or edit what already exists? Are you working with the web designers and developers to create the site? Do you have to create the HTML code or write it in Microsoft Word and let someone else do the HTML? How many web pages are there?

- For help, can you duplicate content from the user guide, or must the content be unique? Do you need context-sensitive or page-sensitive help, or is this something that can wait?
- Do the components need to be delivered in HTML, PDF, hard copy or on DVD? In video or interactive tutorial format?

Chapter 12, "You Want it *How?*" talks more about the different presentation methods you can use for technical documentation.

What Corners Can You Cut?

OK, most of us don't like to talk about "cutting corners," so let's say instead "setting expectations." Once you and the stakeholders have determined the set of documentation needed, break each deliverable down into a "good, better, best" progression. For example, for the how-to help, *good* might be "12 topics done in HTML, covering account questions, billing questions, basic usage, and mobile devices." *Better* could be 30 topics with a new "getting started" section, and *best* might be 48 topics with multimedia tutorials and context-sensitive help.

After you have your priorities set and your good-better-best list in place, refine your calculations: is it more important, to do the "best" job on priority 1, or to do a "good" job on all of the deliverables? Work with your internal stakeholders to determine the minimum requirements for the first release. You don't have to make a matrix of the configuration, but doing so can be helpful.

When you meet with the stakeholders and present your list, you may discover that once they actually see your table, they realize that some things are not important after all. You all may come to the conclusion that troubleshooting isn't so important right now because you don't have enough information, and you don't need a user guide because the online help covers all user needs. It helps to see the deliverables list laid out. I've listed all of the deliverables due for the project, and the level of quality needed.

When time to market is the most critical factor, there is less time for document development during a single release cycle, and it is often expected that the documentation will be refined and enhanced during later releases.

If you add up all the hours spent on research, learning the product, meetings, writing, and production across several releases, you are likely to discover that you spend more hours than you realize on documentation—it just might be that because of rapid releases, your work spans a number of releases rather than being done all at once.

The matrix may look something like this:

HomePrize Cloud Storage 1.0 Deliverables		
Deliverable	*Format*	*Level*
Web page content	HTML	Best
Help for mobile device and Web	HTML, .chm	Better
Troubleshooting	HTML	Postpone
Quick start	Flash	Good
User guide	PDF	Postpone
Developer guide	PDF	Good
Mobile device content	text	Better

This matrix shows that the critical deliverables for the first release are web page content that must be of very high quality, help topics of pretty good quality, and an acceptable quick start and developer guide. As the team looks at the list, all agree that the troubleshooting and user guides can wait until a later release.

You should also re-determine the priorities so you know what you should work on if you have spare time—do you try to create a troubleshooting or user guide? Or improve the quality of the help? (Spare time? Someone has spare time?)

Calculating Time

There are four factors that affect how long any project will take: size, scope, and quality, plus the amount of time required for production. Until you have enough experience to use your own work history to calculate time, follow this rule of thumb for calculating the amount of time you'll need:

Size x Scope x Quality + Production = Hours to Complete

This is no scientific formula; it's a modification of one I've had around a long time, updated for today's tighter schedules. It's just a starting point for those of you who want something to get you going. Increase your estimates until you get some experience. Keep track of your productivity and continue to adjust until the formula matches your reality. (On the other hand, some writers believe that two pages a day is the best way to estimate their time—that could work as well for you as the more complicated formula!)

Size

Plug in the estimated number of pages (for a book or website or multimedia piece) or help topics (for a help library) you expect you'll need to produce. How do you know this number before you've written them?

If you haven't done an outline yet, use similar documents in your company or other companies to give yourself a good idea. If your product has more features or is more complicated than your comparison product, bump up your page count. If the project you are working on is an update of an existing project, calculate the amount of new material being added.

Scope

To estimate scope, there are a few things to take into account. If this is a new product that requires you to do a lot of research and learning, that is a large effort. You'll need to do research, go to meetings, learn the product, and take screen shots or do illustrations as well as writing. If you are updating an existing project, you must assess the amount of rewriting and revision that needs to be done as well as how much information should be removed.

Also think about how important this project is—is it your company's bread-and-butter product or is it for an important high-paying customer, therefore requiring you to spend extra time ensuring a high level of quality? Assign one of five points between **1** and **3** that corresponds most closely with your assessment of the scope of your project:

1 Lots of content reuse from other sources. Little original writing.

1.5 Some copy/paste, mostly straightforward writing on a product you know. No developer input needed.

2 Requires concentration, although not enormously complex.Some developer input required. Source material and design documentation available.

2.5 Much new material and steep learning curve.

3 All new material and a complicated product. Steep learning curve and requires major input from subject matter experts.

Quality

While it would be nice if all deliverables could be at the "best" level, it's not always possible to do everything you need to do to achieve this quality. Assign a number of 1 or 2, depending on the level of polish and validation required:

1 Standard amount of review and revision, with possibly one review cycle.

2 You and others are going to spend time working on it. It will require two or more review cycles and might also require input from many reviewers.

Production

You'll need to add the hours it takes to format and do production tasks on a project. You might calculate 0 additional hours for a job that requires little or no formatting. This can include ASCII text files, wiki content, Word output that someone else will format, simple HTML, or micro-blogging.

If the project is an update and not a brand-new document, don't short-change the production time. You may have calculated only 10 new topics' worth of content in a 40-topic help set, for example, so you might figure that the development work will take only 15 hours. However, the amount of time spent building the help, reviewing help links, and testing the links is the same for 10 topics as it is for the entire 40 topics.

For a job that requires substantial production work, estimate how many hours this takes. Factor in the final checklist activities discussed in Chapter 17, "Wrapping it Up." Production work includes jobs like layouts for designed and printed material, FrameMaker-to-PDF conversions, building help, and creating Flash or other multimedia. If you need to send out the job to a subcontractor, find out how much time the vendor needs after you hand off the content.

Hours to complete

Now multiply your numbers. Size x Scope x Quality + Production Hours will give you an estimate of the number of hours and therefore determine when you can deliver the job. Using this formula, let's look at the calculations on a few of our HomePrize Cloud Storage 1.0 deliverables:

- **Help.** There are 30 help topics. You are familiar with the content and product and assign it a scope of 1.5. Quality rates a 1 because although this is an important new consumer product, you don't expect or need reviews. Production is minimal because you are not responsible for the style sheet or other formatting, and in this case will hand off unformatted text to a web developer. 30 x 1.5 x 1 + 0 = 45 hours.
- **Quick-start guide.** There are about 30 screens in the quick-start Flash tutorial. You are somewhat familiar with the content and assign it a 2. Quality rates a 2 because there are many moving parts that require testing and review. An outside designer requires a week to provide a finished Flash module, which adds 40 hours to the production time. 30 x 2 x 2 = 120 hours + another week of production time.
- **Developer guide.** You estimate 60 pages for the developer guide. You are unfamiliar with the content and expect to have a steep learning curve, so assign it a scope of 2.5. Quality is 2, with many reviews expected. Production time, including building the PDF, is about 8 hours. 60 x 2.5 x 2 + 8 = 300 hours + one day of production time.

Working Backward to Move Forward

To plan a schedule for a due date in the future, simply work backward from your target deadline. Find your deadline on the calendar, then calculate the amount of time you have to do everything between your due date and today's date. To begin, you'll figure out the milestones you need for your document and indicate the only dates you know right now—today's date and the final due date. For example:

August						
Su	M	Tu	W	Th	F	Sa
				1	2	3
4	5	6	7	8	9	10
11	12	13	14	15	16	17
18	19	20	21	22	23	24
25	26	27	28	29	30	31

1. Today: **August 1**
2. First draft sent out: **??**
3. Review comments from draft 1 returned: **??**
4. Second draft goes out: **??**
5. Review comments from draft 2 returned: **??**
6. Electronic documents posted to extranet: **August 30**.

Now, reverse the chronological order and put your final date at the top. Look at a calendar as you plan the dates so you have a visual aid to see the number of days between the milestones. (You don't have to actually reverse your calendar as I have below, but do think "backward.") Be aware of weekends!

August						
Sa	F	Th	W	Tu	M	Su
31	30	29	28	27	26	25
24	23	22	21	20	19	18
17	16	15	14	13	12	11
10	9	8	7	6	5	4
3	2	1				

Your thought process will go something like this as you start at the end date and work your way back to today:

6. **Electronic documents posted to extranet.** August 30. *That date is fixed.*

5. **Review comments from draft 2 returned.** *How much time do I need between this step and August 30 to incorporate the final review comments and do my document production? I think I can do it in three days, since it will be my second time incorporating feedback. I'll say the 27th.*

4. **Second draft goes out to reviewers.** *Reviewers would like five days to read and comment on a document of this size, but of course they've already seen it once so maybe they don't need the full five days. I'll send out the draft on the 21st, which gives them nearly a full week if they use the weekend. I can use that time to do cleanup.*

3. **Review comments from draft 1 returned.** *After I get the review comments, I'll need five days or more to incorporate all the feedback. Five business days before the 21st is the 14th. If it's more work than anticipated, I'll have the weekend to work.*

2. **First draft sent out.** *I'd like to give reviewers at least five days to look at this document. Since I'm asking them to return it on the 14th, I would need to finish the first draft by the 7th. That's really tight because today is the 1st. I'll send my draft out on the 8th and continue to work on the document while it's out for the first review. I'll just have to tell people which sections are to be supplied (TBS).*

1. **Today.** August 1.

As you work your way back to the starting point, you may find that you have to make adjustments. For example, in the scenario above, you might decide that you can't possibly do a first draft in seven days and you need at least two weeks to do it. If that's the case, make the two-week date your starting point and work backward from your end date. You can always adjust the schedule by removing days or removing a milestone.

And that's how you plan a schedule. You see now why those milestones discussed in Chapter 9, "Process and Planning," are so important and why it can be so useful to write up a thorough document plan.

We've covered a lot in this chapter, and don't feel bad if you don't get it right away. There are a lot of factors involved in estimating how to get a job done, and many of them have nothing to do with you—releases are delayed, new features are added, priorities shift, and people quit. But the more you know and the more you can estimate in advance, the more control you'll have over your work life.

Chapter 12

You Want it *How?*

Tools and technology: what you need to know and do to bring your content to life.

What's in this chapter

- The tools and technologies a tech writer should master
- Where to store your content
- What to think about when planning books, web content, social media, and online help
- Why tool knowledge can help you be a better technical writer

When we went over all the different types of documentation in Chapter 8, "The Deliverables," we didn't really discuss what this documentation might look like—whether it was an animated video, a book, an HTML page or help text on a website or a mobile device. It could be any of those—in fact, with single sourcing, a piece of content could be *all* of those. I didn't want to focus too much on the form the documentation might take, because it is more important to first think about the *type* of content you need to produce, and then figure out how to produce it.

These days, tech writer equals content developer and you'll be expected to produce content in many different formats for every possible way a user may want to access it. Today's tech writer often has to know how to do design and layout, take photographs, do illustrations, and build help in addition to writing and organizing content. You'll need to know all about the best tools to use to develop content for many different types of output.

This chapter will acquaint you with some of the many ways you might produce documentation and help you understand why you might choose a specific format. I'll also discuss some of the tools and technologies that you can use to create any kind of deliverable you want.

If I don't mention a product here, it's not because it's bad; it's because there are a lot of choices and limited space to discuss them. Besides, software preferences change often, which can mean that the newest and best product on today's market may not even be available a few years from the time of this writing.

Today's Tech Writer Carries a Big Toolbox

It pays to have as many tools in your toolbox as possible. Today's technical writer should know how to use the right tools to deliver all of the following:

- **Structured content with XML.** Today's technical writer should be familiar with structured authoring with XML. *Structured content* refers to content that has an organizational structure forced upon it. XML content can be published to a book, web page content, help, or other form of output. This takes advantage of "single sourcing"—the use of a single set of content to be released in multiple formats. This will let you deliver your content in the right form to the right audience, and that's ultimately what your job is all about.
- **Books (manuals), pamphlets, and data sheets** in both print and PDF or on an e-reader device. This means also knowing how to do screen shots, image editing, and simple illustration.
- **Web page content.** This means knowing not only HTML, but also having some knowledge of XML, CSS, and JavaScript. You may also be creating content for a knowledge base, a wiki, a blog, an e-commerce site, or web help.
- **Help.** This means knowing how to design, write, and generate help for users of software, the Web, and mobile devices. Your help can be made of anything from simple HTML or XML files, to wizards or complicated interactive Flash tutorials and multimedia output.

That's a big skill set for one tech writer to have! Luckily, you don't have to learn all of these skills at once. And often, you can learn them on the job.

A tech writer with something extra may also be called upon to understand a programming language such as C++ and Java, or UNIX or Linux command line syntax.

Location, Location, Location

Whether you produce mostly PDFs, mostly online help, or a combination of the two, someone must make a decision about where the content will reside. This means that it must be available to both internal and external customers, and it also must be stored for later use by the technical writing team. Many software products include help and PDF documentation in the software itself—in the "build" (the process of converting code into a standalone product). When the software is "built," the content is included, as part of the software.

Including documentation within the build does guarantee that the help is always available, and not at the mercy of system maintenance or network problems. It can also be faster to access. Some people also believe that it's important to include help in the build for users who don't have Internet access, although I'm not sure how many people today use software and don't connect to the Internet.

One down side to including documentation within the build is that you cannot change it until a new version is released. If you often put out document revisions, then including documentation within the build may not be for you.

The other down side to putting help in the build is that it can take up a lot space. Consumer electronics, desktop software applications, and mobile applications can't afford unlimited space, so their help content is often stored on a *web server* (a computer that "serves up" web content to a browser). When a user clicks the help icon in the software's user interface (UI), a browser displays help that actually resides on the server.

Externally Facing

Many companies put a collection of PDF and web-based documentation on their corporate website, where customers (and anybody) can read and download it freely. This is convenient for customers, certainly, as long as they can find it easily. Often, it's linked from the software's user interface.

For this to work, the files must reside someplace where the URL will not change. You need to be able to freely add content to this location so you can update it. This is not possible on all company websites. Even if your company is willing to put technical documentation on the corporate website, policy might prevent *you* from pushing content. Or, the site might have dynamic URLs so there is no guarantee that a link from the application will be stable from one day to the next.

One solution that many companies use is to create an *extranet* site for customers. An extranet is a private network that uses Internet technology to share part of a company's content with users who must be granted permission to enter. Customers access the newest documentation by logging into the extranet site. You can manage your content on this site. And best of all, you can create a link from the software's user interface to the extranet site.

A hybrid approach can be the best bet—including help files in the build when you can, public documentation on the corporate website, and more proprietary documentation on an extranet. This takes some planning and maintenance.

Internally Facing

Some thought must be given to where the documentation resides *within* your organization. You need a place for your working files, which should be stored in a place where nobody outside of Tech Pubs can get to them. You also need a place for finished documentation, where your internal customers can find what they need.

A lightweight content management site such as SharePoint can do double—and even triple—duty for document storage and accessibility. Its version control system lets you check files in and out to prevent overwriting of files so you can store all of your source files in it and check them out to work on them. It can be set up as an intranet, extranet, or even an Internet site, allowing appropriately tagged files to be viewed and accessed by internal and external customers.

Internal storage

Some Tech Pubs departments store documentation in and work from the revision control system used by the software developers, such as Subversion (SVN).

Revision control helps in the management of files being worked on by more than one person. To work on a file, you must check it out, and then check it in when you're done. If two people work on a file at the same time, the last one to do so must either merge changes or overwrite changes. If necessary, you can also roll back to a previous version. Discuss revision control with the development team to see if it can be a solution for your Tech Pubs department.

Internal distribution

To make documentation available to your internal customers, I recommend an internal website. It should be part of your final handoff to post finished documents to this site, so these documents become available to everyone in the company who needs them. This can be anything from a wiki site, to a homemade site created and maintained by your department, to a SharePoint site. If you have the option, I would recommend a wiki, but the tool you use to create this site is less important than having it and maintaining it consistently.

If no such website exists, volunteer to create one, and then make sure everyone in the company knows about it!

The Right Tool for the Right Job

Most experienced writers who have used a range of tech-writing tools, can learn any new tool quickly. After all, the most important tool at your disposal is your brain, and you can use it to learn what you need to learn.

But employers are funny that way—they like to hire people who have expertise with their applications. So it's not a bad idea to have the right keywords on your resume, and the ability to talk intelligently about your expertise during an interview. Your knowledge of the tools of the trade helps you remain employable, which is no small thing in today's uncertain work world.

And the more you know, the easier it is to learn more. If you are an expert user in one type of help authoring tool, it's relatively easy to learn another one. It also equips you to choose the right application to do the best job on the content you want to create. Making recommendations will often be part of your job.

XML

As you might guess from the number of times it's mentioned in this book, XML is growing more and more popular as a means for developing and managing content. Many companies are turning to XML in an attempt to streamline documentation production and content reuse, so it is a good idea to understand XML if you are to work as a technical communicator.

What is XML? XML is a flexible meta-language (meaning a language that lets us create or define other languages!). That's a mouthful, so let me try to explain it. XML allows you to define tags (similar in appearance to HTML tags) that can be used to mark up content. It is used for *structured authoring*, that is, writing that follows the enforcement of organizational structure of content components. (There are types of structured content other than XML, but it is very unlikely you will be using any other.)

For example, structure rules might determine that all bullet lists must be preceded by an introductory component. Or that all help topics must include a list of cross-references at the end. Or that a certain component may not be nested within another.

In practice, even though XML allows tags to be defined by the user, you will almost surely not define your own tags. Instead, you will use a pre-defined schema or DTD, which is a file or set of files that define the tags and structure. Nearly all organizations these days use one of a few standard XML schemas for technical communication. The two you will hear about most often are DITA and DocBook. While these are the most common, you may discover that your company uses a proprietary schema or an industry specific schema like S1000D.

The XML schemas you're mostly likely to work with, like DocBook and DITA, do not use tags to define the way content looks, for example, what color or size an element is. In HTML, the <i> tag renders the contents in an italic typeface, which might be used for emphasis, a programming variable, or something else. However, the <i> tag only specifies the typeface, and says nothing about the content.

In XML, content like emphasis or programming variables could be given tags that indicate their content: a component that is a warning message would always have a tag that indicates it's a warning. A component that contains price content would always be tagged to show it's a price. These tags then are used to control the order, processing, or structure of the content. The main advantage of XML is that it separates content from presentation.

Separating Content from Presentation

Once all of the content is created, the assembled content components are output it in any of a number of formats. Because XML is unformatted content, it can be sent easily to wherever it needs to go—through the Internet via *web services* (a machine-to-machine interaction over a network), to a publishing tool like Adobe FrameMaker, to HTML online help file formatting, and more.

Be aware that your customers may use different operating systems or browsers. "Cross-platform" capability is key when delivering documentation, and XML, because it is not tied to any specific platform, allows you to deliver that capability.

This enables a Tech Pubs department to automate content layout. Instead of making daily decisions about which content components go where, how they look, and where the pages break (while continually tweaking and rearranging), a writer working in a structured authoring environment simply focuses on content. This can save a lot of time for a company that produces a lot of documentation.

And it can add much-needed consistency, in content and structure. Reused content reduces the likelihood of using inconsistent terminology. It also means that each document of a certain type contains components in the same order and abiding by the same rules.

To work in a structured authoring environment, you will develop content in an XML editor. Although XML content can be created in almost any application, it's best to work in an application, like PTC Arbortext or XMetaL, that enforces your structure. Adobe FrameMaker comes in both structured and unstructured (just be aware that you cannot move freely back and forth between the two). How the content looks will be managed by a style sheet like CSS or any of a number of publishing output tools that can be used to apply formatting to the final output.

Making XML Work

If your organization does decide to use XML, you are going to want to look into either DITA (Darwin Information Typing Architecture) or DocBook. You can use nearly any XML editor to produce DITA or DocBook content.

DITA is an XML-based open standard for authoring and publishing content. Its information architecture is based on three topic types—task, concept, and reference—and you develop your content in these topics. DITA maps assemble the topics to produce output. DITA is distinguished by being easily customizable through a process called specialization.

The practical result of this for writers is that when you work in a DITA shop, you will probably be working with unique elements that have been created for the environment you are working in. DITA has a reputation for complexity, but when well implemented, the complexity should be hidden from writers. Learn more about DITA at **ibm.com/developerworks/xml/library/x-dita**.

DocBook is an XML schema that has been around since before the days of XML. It was originated by a group of companies that wanted to share software documentation. It has been adopted by many software companies, especially open source projects. It uses more traditional structures than DITA, with elements for books, articles, and all of the things that you would expect inside. Learn more about DocBook at **docbook.org**.

DITA is often considered to be better than DocBook for content reuse and topic-based authoring. DocBook is often used for linear books, although it can handle topic-based authoring as well.

If you are working in a company that does not currently use XML and structured authoring and you are in a decision-making capacity about going forward, make sure you allow plenty of time both to make the decision and then to implement it. Depending on how much documentation your organization produces, implementing structured authoring and converting existing document files can be expensive and take quite a bit of time, sometimes up to a couple of years with the help of many consultants.

If you don't produce a lot of documentation or don't have significant need for reuse within your department and throughout the company, you may decide that it is not worth the investment.

Go Green—Reuse Your Content

Content reuse is an important part of structured authoring. The content must be in small enough chunks to easily reuse without applying excessive *conditional text* (content that is marked so it appears in some output but not others) and tagging. A content chunk can range anywhere from a few words to an entire long section to an image to an animated video to...well, just about anything.

You can't just dump a bunch of content in a directory and hope technical writers can figure out how to use it. Storing content for reuse requires rigorous maintenance so that any piece of content has exactly one location and can be easily located by the writers. Edits are made *only* to a single file (called the source file.) and that file is linked to from the output documentation. You need plans, outlines, document maps, README files, and whatever it takes to help writers understand how to work with the content to create output.

Conditional text can be used in both structured and unstructured content.

By turning various conditions on and off, you can produce a lot of customized content from a single file. With conditions, you can change the look and content in many ways, by adding and subtracting content, eliminating or adding images, changing product names, and much more.

You need an overall plan—for creating and maintaining content chunks, for storing and organizing your content, for setting structural rules for deliverables. You also will have to plan each project.

You also can't take the same content source and simply repurpose it. For example, there is not a lot of value in writing a user guide and then converting the entire user guide into help. Users often refer to the help and then the user guide when they can't find what they want. Or they refer to the user guide and then check the help for more information. They aren't happy if they find the exact same information in both. Instead, use content that makes sense for each type of deliverable, add more when necessary, and remove any that's not necessary.

Content Reuse Without XML

Many people think that XML, or structured, content is required for content reuse, but the fact is that you can also use unstructured content such as that produced in FrameMaker or Word.

- You can write linear documentation, like a book, and then use conditional text to create a subset of the content to produce other deliverables such as a brochure or help for a website and a mobile device. Building a document in

Adobe FrameMaker or another desktop publishing tool and generating help from it with Adobe RoboHelp or WebWorks ePublisher or Madcap Flare is one way to do this.

- You can write help based on shorter, nonlinear topics, and then add more content to create a book or other deliverable. Madcap Flare is an example of one XML-based authoring tool that you can use to write help topics and assemble them into a manual.
- Or you can create many standalone topics with no predetermination of where they will go (XML is best for this, but as I said, you can use any of your standard tools), then mix and match them later.

Whether you use XML or another form of content, that content has to be presented to customers in some sort of format. Content destined for reuse can be presented in many ways, using different tools.

In fact, all content can be presented in various ways and there are tools to help you produce virtually everything you can think of. The rest of this chapter talks about some of the output you can produce and how you can make it happen.

Books and Other Narrative Material

Today, a book can be many things. It can be a set of pages between two covers, something you hold in your hands and turn the pages of as you read it. It can be an electronic deliverable in PDF form. It may also be an ebook, designed for any of a number of e-reader devices. In all of these cases, though, a book is made up of multiple sections (chapters, normally), has a unifying design, or layout, and can be read in a linear, or narrative, fashion from beginning to end to provide cohesive information.

Building with Desktop Publishing Tools

Whether your book is printed hard copy or PDF, it needs to have a layout, and requires a desktop publishing or comparable tool that lets you bring all of the elements together.

Adobe FrameMaker is a favorite of tech writers everywhere for developing multi-file documents. FrameMaker is a very stable program that works well with large books, and it uses templates, which enable you to produce multiple documents with the same formatting.

FrameMaker can produce either unstructured (tagged content), structured, or XML content. FrameMaker also works with several help authoring tools: Adobe RoboHelp, and WebWorks ePublisher, in particular. As part of Adobe's Technical Communication Suite, FrameMaker is integrated with RoboHelp.

FrameMaker does not have many competitors when it comes to producing multi-chapter technical documents. But it's not the only game in town, and there are other ways to produce books. You might use publishing applications like QuarkXpress, Corel Ventura, or Microsoft Word, or an open source tool such as Scribus.

Whatever the tool, you need a set of good templates to develop multi-file publications. You'll find more about templates in Chapter 20, "Design and Layout."

Whatever happened to those nice printed manuals?

Every now and then I hear someone lament the demise of printed documentation. As a matter of fact, printed books still exist, but they are in your bookstore, and you have to pay for them. There are many books available for the most popular products—take a look at the Adobe Classroom in a Book series, for example—but the fact remains that you are unlikely to have hard-copy documentation with software. It's just too expensive to produce.

It isn't just the printing costs, but the labor, shipping, and materials that cost so much (In the old days, companies often put documentation in binders so they could reprint single pages with updates and corrections. Those pages would be inserted in the existing binder or mailed to the customer.) Rapid changes and numerous releases make it nearly impossible to keep up with a printing schedule. It makes sense to produce documentation in knowledge base form (relatively easy to update), PDF (less easy, but much cheaper to change than the old-style printed book), or HTML (very easy if you use a wiki, which can be changed on the fly).

If your organization does a lot of printing, look for a printing firm that will provide you with an account representative who understands your needs. Typically, you will send a PDF of your ready-to-print document to the printing firm and you won't see it again until you pick up the finished product. However, I recommend that you review a proof copy to make sure that the color looks the way you want it to, the pages are trimmed the way you expect, and the paper (*stock*, as they call it in the printing world) is what you want.

While price will of course be a consideration in your choice of vendor, having a reliable contact at a firm you trust is, as they say, "priceless."

There are companies that still produce printed documentation, and your company may be one, particularly if you sell off-the-shelf consumer products. As I mentioned earlier, printed documentation is common with consumer products such as electronic devices or appliances.

When do you want to provide printed documentation? When you want a good out-of-the-box experience for your customers. When customers need to get started immediately with no time wasted. When any difficulty using the product adds to the risk that the customer will return the item. Or when the customer has no immediate access to the Internet for documentation but still needs to install. (If you are writing documentation for a device such as a computer, modem, or wireless router, you must assume users have no Internet access until after the device has been installed.)

PDF

PDFs provide the same kind of laid-out, designed books that you find in a printed hard copy, except that the users are expected to do the printing themselves. Using the Internet as a delivery mechanism has allowed companies to save huge amounts of money by not having to buy printed manuals in quantity. They were often thrown away when a product didn't sell as much as expected or when a new release came out and made the old one obsolete.

A PDF offers great convenience to the users as well. Anyone who installs a free reader can look at PDFs. They can open the document on screen and zoom in or out to make the font size whatever they want. Because it costs no more to produce color than black and white, a PDF can be done in full color, which can be both useful and attractive. And most importantly, a PDF is *cross-platform*, which means it can be opened and read on any operating system.

To create PDFs, you use the same publishing tools you use to create printed materials. Then you need a program like Adobe Acrobat or PDFCreator to convert the document to a PDF.

Ebooks

Besides PDF, there are many other formats for ebooks. ePub format, an open standard used for ebook content, is one you should look into if you want to create an ebook. If the content is more important than the layout, a format like ePub is preferable to PDF—it's what lets content flow when a user adjusts type size, for example. ePub lets your ebook be read on just about any ebook device.

Many applications export to ePub files, including Adobe InDesign. If you use FrameMaker for book and help content, you can export FrameMaker content to ePub by using RoboHelp (it comes with FrameMaker in the Adobe Technical Communication Suite) or WebWorks ePublisher. On the Mac, the Pages office software can export to ePub.

calibre is an open source e-book library management application that also converts documents to ePub and other ebook formats. Find out more at the calibre website **calibre-ebook.com**

Pamphlets and brochures

Shorter pieces such as data sheets or brochures are often used for marketing and sometimes require fancier design tricks than a word processing or desktop publishing tool can handle. If you find yourself needing to include a lot of graphics, text wrapping, or unusual columnar layout, try an application intended for design such as Adobe InDesign or QuarkXpress.

A very practical reason for avoiding excessive screen shots is that last-minute changes to the user interface can mean that you have to go in and reshoot screen shots. Minimize the possibility of this happening by using only screen shots that provide useful information.

When you do use a screen shot, there is no need to show areas that are not relevant, unless they are required for giving the user context.

Graphics Illustrate Your Point

Yes, a picture can be worth a thousand words. A good visual can tremendously increase the information value of a document. Often, an illustration can explain something clearly that is very hard to describe in words.

People have different types of learning styles. The three most common are visual, auditory, and kinesthetic. In a nutshell, visual learners benefit from illustrations, auditory learners read aloud or otherwise verbalize to themselves, and kinesthetic learners learn by doing. (Good classroom training incorporates all of these styles at once, with a speaker, slide presentations, and hands-on practice.) Adding illustrations to your documentation lets you appeal to visual learners.

Understanding Graphic Formats

The types of graphics you'll use to illustrate your documentation come in two flavors: bitmap and vector.

Bitmap files

Bitmaps, or raster files, are made up of pixels. Most of the images you see on your computer, including all the screen shots you'll take, are bitmaps. Image files with the extension .jpg, .tif, .bmp, or .gif are bitmap files. If you aren't sure what file format to use, use .jpg for photographs and .gif or .png for screenshots and other nonphotographic uses.

Bitmap files can be scaled down nicely, but often lose some quality when enlarged. However, the larger the image, the larger the file size. You might discover that a PDF with a lot of high-resolution screen shots in it can be absolutely huge.

Control file size by making sure images are sized appropriately before being imported into a document file.

Bitmap files can be edited only by an image-editing program such as Adobe Photoshop. Image-editing programs allow you to use *layers*, which let you add text and drawing shapes to the image. Before you can use the graphic, it must be flattened to turn it into a .gif or .jpg. If you do any image editing of this sort, make sure you keep the layered files in addition to the output. I guarantee you will need the source files again, especially if your company does translations. (I actually recommend against adding text to a bitmap. Learn more about that in Chapter 21, "Gaining a Global Perspective: Localization and Translation.")

Vector files

Unlike bitmap files, vector files are made up not of pixels, but of paths, which can be lines or shapes. These files do not lose quality no matter the size. Files with the extension .ai, .eps, .svg, and .drw are vector files.

Vector files can be converted to bitmaps, but the quality will be degraded. Keep your illustrations in vector formats and import them into your publishing tools in their native formats.

Vector files are used for line illustrations. Whether you are doing a simple line drawing within FrameMaker or Flash, or a complicated illustration in Adobe Illustrator, you are creating a vector file. I recommend using the .svg (Scalable-Vector Graphic) format for vector drawings. svg is an XML-based standard. An .svg file can be imported into publishing programs such as Adobe FrameMaker and displayed in browsers. It's a good format choice for illustrations if your organization does translations, since the content can be easily extracted for that purpose.

INSIDERS KNOW

It's a good idea for all content contributors to create screen shots and other images in the same format and the same size. This avoids surprises when single-sourcing. Often writers manually reduce images when importing them into a book, only to be surprised when the same images are huge when displayed in a browser.)

Screen Shots

Screen shots, or screen captures, are very common in user documentation. A screen shot is a picture taken of part or all of your computer screen, typically showing a key part of the user interface. As the user works through the tasks in a procedure, he can compare what's on screen with what is in the documentation.

Don't let your screen shots be too much of a good thing, though. Excessive screen shots are not particularly helpful, and some technical writers fill page after page with screen shots that add little or no information and take up so much space, the procedural steps are lost. A user can see what's on the screen and doesn't need to have dozens of pictures to confirm it.

A really useful screen shot is one that shows the user what to enter into a field, or shows the user which button or other on-screen object is being discussed. *Callouts*, or captions with pointers, draw the user's attention to the important areas.

There are many good inexpensive or free applications for taking screen shots, including those that come with your operating system. I suggest that you look for features that will be useful to you, like TechSmith Snagit's scrolling feature that captures even the unseen part of a web page, or the full-featured image-editing functionality of Corel Paint Shop Pro.

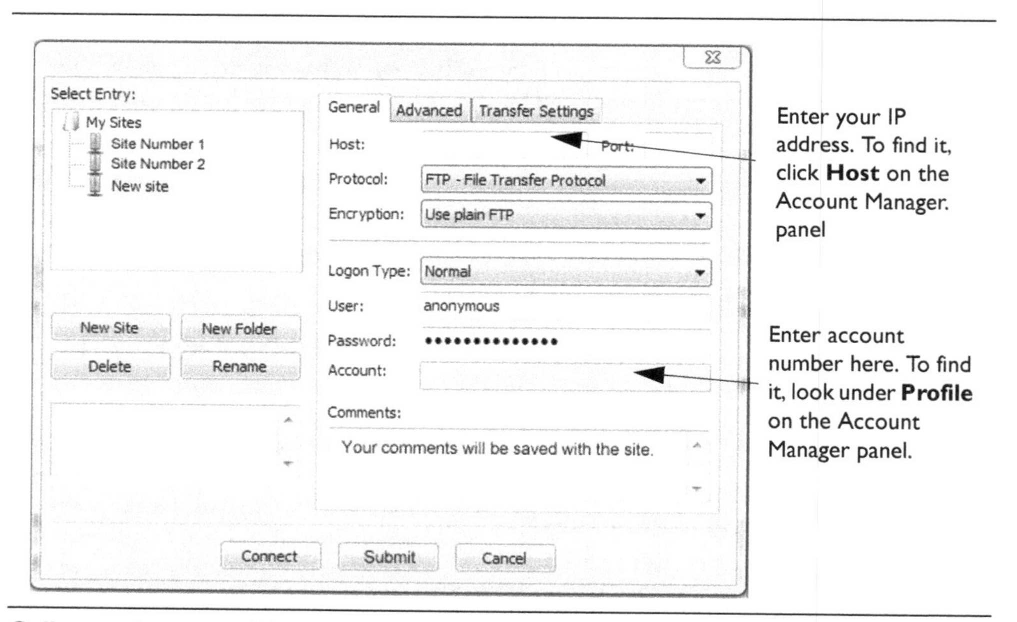

Callouts, the text with arrows, add value to a screen shot by providing specific information that can help users instead of just showing them what they already can see for themselves.

Image editing

In many companies, technical writers end up taking most of the photographs used in technical documentation. Sometimes there is only one chance to take the picture before the prototype or device leaves the premises, and if the picture has a problem, you can't reshoot. You must use photo-retouch skills to fix any problems that occur.

Even if you are not the house photographer, you'll discover there are many occasions when you will need to manipulate images. For example, you may take a screen shot that has actual customer data or employee data in it. You'll need to replace that data with neutral information that can be used for an example. Or maybe your documentation is ready to go when you discover that some features were rearranged at the last minute. It is sometimes faster to edit the image than to go back to the application, set up the data, and take new screen shots.

Adobe Photoshop is probably the best-known application for this purpose, and it's a good one to know and to have on your resume, although likely to be more than you'll need for most of your purposes. Investigate alternatives such as Corel PaintShop Pro, Gimp, or Paint.NET. Look around and you're sure to find one you like.

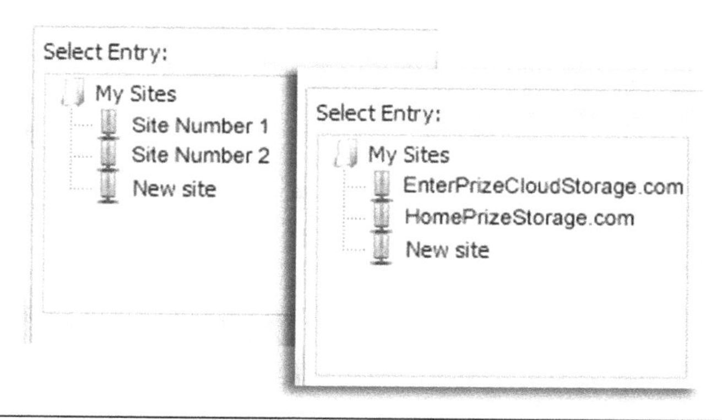

You don't have to import data into an application or take new screen shots when you want to show more appropriate data. The revised portion of the screen shot on the right was made by using the text feature in an image-editing program.

Illustration

Very few technical writers are also good technical illustrators (and vice-versa!), and I wouldn't suggest you do much more than simple drawings, leaving the complex technical illustrations to the professionals. With a program such as Adobe Illustrator, or the drawing tools in FrameMaker or Microsoft PowerPoint, you can create useful illustrations, including charts and graphs. You can also use these tools to modify illustrations that someone else has done, which is critical as features and functionality change with each release.

A diagram can act as a quick reference for a long, complicated procedure. The following figure shows a flowchart of an actual procedure that had taken nine pages to explain. A diagram like this can summarize the process in less than half a page—a huge saving of space and text. A user can scan the diagram for a quick understanding, or refer to it while reading the description.

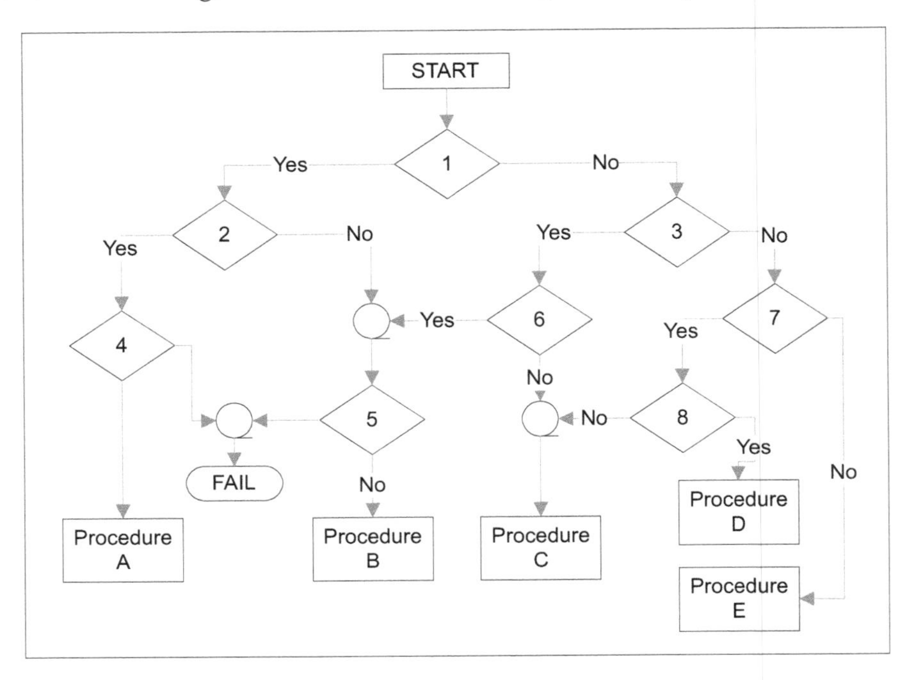

Microsoft Visio is probably the most common tool used to create diagrams and flowcharts. You might also want to investigate Dia, the open-source alternative.

Writing for the Web

If the product you document is a web-based application or e-commerce site, it's likely that your main documentation tool is an HTML editor. You may also find that you need to be expert in CSS (Cascading Style Sheets), a style sheet language that controls layout, colors, and fonts in web pages.

And there is a host of other technologies that drive web applications like JavaScript, XML, and Ajax, but when starting out, just make sure you can edit HTML without breaking anything. Choose a WYSIWYG (pronounced "wiz-eewig" for "What you see is what you get") HTML editor that doesn't require you to code from scratch, like Adobe Dreamweaver or Microsoft Expression Web.

Single-sourcing can also come into play, since you can convert FrameMaker, Word, and other sources into HTML. Just make sure that the output is acceptable to the web developers in your company. These types of conversions do not produce very clean code, and don't work well with existing CSS. But for internal use or as prototypes, they should be fine.

Knowledge Base

A knowledge base is a database of content that resides on your server and is designed to answer questions and solve problems for users of your company's products and solutions. Knowledge bases typically have two purposes—to allow people *within* an organization to access information, and more important, to enable *external* customers to access information from the company's website. Perhaps you've heard about "self-help"—a system of tools and information that allow a customer to answer questions and solve problems without having to call Customer Support. Many customers like being able to help themselves, and most companies love cutting down on support calls, which can be very expensive.

If there is a movement toward self-help in your organization, make sure to get involved. You want to make sure that customers do not receive conflicting information, so it's important that your product documentation and the knowledge base are in sync.

A knowledge base can be made up of product documentation, short topics, even emails written from one employee to another. The users can search for any topic on which they want answers. Some knowledge bases have built-in intelligence, meaning they refine their responses based on actions the customer takes.

Anne Gentle's book, *Conversation and Community: The Social Web for Documentation*, is about how technical writers can foster conversations and community using social media. Take a look at this book if you too see your role as one of sharing information through blogs, wikis, micro-blogging, and syndication.

Social Media

I've talked a lot about developers who are too busy to review the documentation, but what if you had a group of people at your beck and call who wanted to know about your product and were eager to help you improve your documentation? Sound like a dream?

Well, it may be more available than you think. We've already discussed how technical writers should use social media to further their own career interests.

In addition to that, there are growing opportunities to use social media to distribute and improve technical documentation.

You can use an outlet such as Twitter or Facebook to distribute product highlights or release notes for your company. You can use forums, blogs, and social networking sites to engage in continuing real-time conversations with users to learn what they want and think. You, or someone in your company might write a blog, which should be interactive, with feedback from customers. Community forums exist where users help each other use the product. If you are fortunate enough to work for a company that has user communities dedicated to its products, get involved! Not only will you have an opportunity to help people learn about your product, but also you'll get as much input as you can handle about how to improve the documentation.

To play a role in the social media world, you can't be satisfied with the status quo, where you send out a document and forget about it until the next release. Social networking means collaboration and involvement, and if that involvement sounds attractive to you, it can be an enormous opportunity to improve your documentation, help your company, and build your tech-writing career.

Wiki

A wiki is a website that allows collaborative editing and uses a very simplified markup language. Wikis are designed to allow all users to easily edit any page or create new pages within the wiki site. Many companies use wiki implementations like Confluence for their internal documentation, and often a tech writer is responsible for gathering the documentation and managing the site.

Wiki markup languages are easy to learn and use, but in fact, wikis typically use a WYSIWYG editor that doesn't even require the user to learn wiki markup. Try out your wiki skills by contributing to what is perhaps the most famous wiki—Wikipedia—at **wikipedia.org**—or perhaps even better, show off your tech-writing skills at wikiHow the "how-to manual that you can edit" at **wikihow.com**.

Developing Help

When using software, whether it's a desktop application, a browser-based application, or a mobile app, users don't want to read a book to be able to figure out what to do. They just want a product that is intuitive and easy to use without requiring a lot of study. Explanations and help should be no more than a click away, or even closer.

The best help is in context. That means that the information is available at the moment the user needs it, on the spot where the user needs it.

Context-sensitivity, or *page-sensitivity*, refers to help that displays content relevant to the place where the user is at that moment. For example, if you are on the "Add Users" page of a web application and you click a help icon at the top of the page, page-sensitive help displays information about adding a user. Context-sensitive help can be even more targeted, providing help icons at various places on the page. A specific area on a page may have a help icon to bring up content explicitly for that area. Or mousing over that same area can display help content in a balloon.

When you start developing help, you'll discover that there are different output formats, depending on the operating system in which the software runs. For example, PC software products often use help created with the .chm Microsoft proprietary format. If you provide help for desktop software that runs on both a PC and Mac, make sure your help works on both operating systems. (Ask the Mac developers what they need from you.) Help authoring tools like Madcap Flare can produce help output of all kinds, including for mobile devices.

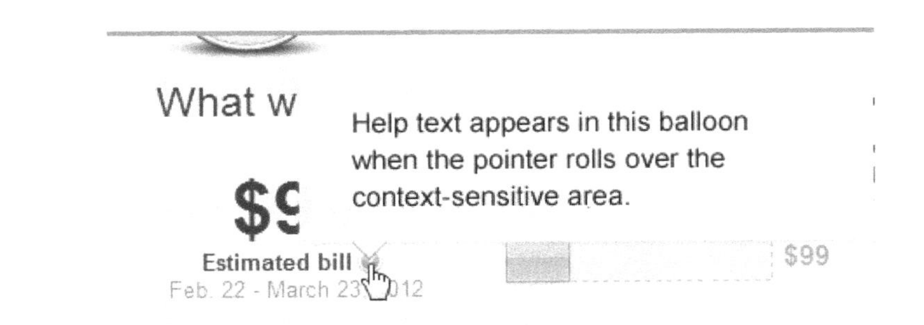

Context-sensitive help displays information in balloons or pop-ups when the user mouses over (rolls over) the area or clicks the help icon.

As with any tech-writing project, developing help requires thought and planning. Some things to consider:

- What operating systems and devices does the product support? Do you need to create help for all of them?
- Will the help reside in the software or be linked to an external site?
- Do you need to produce page-sensitive or context-sensitive help for a website or desktop software?
- What program do you use to generate the source content? Will it be the same application you use to create help, or do you create content in another application like FrameMaker and want to generate help from that?
- What is the structure of the help? What links to what?

When you decide what kind of help you'll be creating, you will then have an idea of which Help Authoring Tools (or HATs, as they are often called) are needed to produce the help. Tools can be anything from an HTML or XML editor to a full HAT such as RoboHelp or MadCap Flare. If you use FrameMaker as your documentation tool, you might use WebWorks ePublisher to build help from it.

Chapter 15, "Putting It All Together," provides some tips on how to plan and build help.

Multimedia

When do you decide to develop a multimedia piece? When standard written documentation is not enough. Perhaps you want to appeal to the kinesthetic or visual user. Perhaps a written explanation just doesn't get the point across as well as animation. Or perhaps this product is so new and unique and important to the business that a visual piece will help not only with user satisfaction but also with marketing.

Decide what your goal is for the piece. For example:

> After watching this tutorial:
>
> The user will be able to connect a new media player and watch a sample video.
>
> *Or:*
>
> The user will understand the key features of the help-authoring software.
>
> *Or:*
>
> The user will be able to understand how to get started with the company's new consumer product, HomePrize Cloud Storage.
>
> *Or:*
>
> The user will be able to back up files in the Documents folder.

Videos or interactive tutorials are often designed for a first-time user, so include as much explanation as you need to. Add pictures where pictures are needed, but don't just plop a screen shot in there with no explanation and expect the user to get anything from it. Use captions, callouts, and animations to point out the places in the screen shot where the user needs to interact.

A tech writer has many options for developing multimedia, including these:

- Applications such as TechSmith Camtasia and Adobe Captivate let you record your own on-screen activity and add captions or even voiceover, then save in a number of formats.

- Recording with a video camera with or without sound can be a good solution for showing how to do anything that doesn't require closeups of a computer screen. You can use video editing software to modify if necessary.
- Adobe Flash lets you build sophisticated animation and interactive pieces. The rich graphic effects you see on many websites are typically done with Flash. Flash is a great tool to learn and will help you produce dazzling effects, but it has a very steep learning curve.
- Microsoft PowerPoint has excellent animation and interactive features.
- You can embed animation in a PDF file as well. Think of the extra zing your PDFs can have if a user can click a link and open an animation.

If you are doing a screen recording, go through the entire process and time it. Keep everything to just a few minutes, if it is mostly informational. (If it is interactive and the user controls the timing, you don't have to worry as much about this.) If it looks as if your recording will be longer, consider making a number of shorter pieces with shorter tasks.

When creating multimedia pieces, you can record on-screen activity and add captions and interaction. Voiceover can be nice, but be cautious—for these reasons:

- To get professional results, you may need a voiceover professional. Your co-worker might sound like James Earl Jones in daily life, but when you listen to a digital recording of him, you might be unpleasantly surprised.
- Professional voiceovers cost money. It will cost more each time you have to have the professional come back to make updates.
- Audio sound tracks are difficult for amateurs to edit.
- A voiceover without captions does not work for the hearing-impaired.
- It's costly and difficult to manage translations for voiceovers.

Test regularly as you work, by going through the entire piece including all interactive actions. If you find yourself becoming bored or feeling as if it's too long, assume your customers will feel that, too, probably long before you do.

Going Above and Beyond the Basics

The advantage to learning these tools and technologies is that you are positioned to make recommendations for improving documentation within your company. If the business is doing things the way it always has, with PDF manuals in book form, maybe you can become a hero and improve customer satisfaction by proposing that you produce help with embedded videos.

If no one in your department is using social networking to gather customer feedback and improve communications, maybe you're the one to launch your organization into today by creating a company Facebook page, Twitter feed, or simply communicating with your users.

Good technical documentation is all about what is best for the users. And by understanding all of your options, you will be able to make the most of today's technology to improve customer satisfaction, and ultimately to improve your own job satisfaction.

Part 4. On the Job

With the material in this section, you could walk into any Tech Pubs department and sit down and do the work. Now you're finally cooking–or at least, writing!

Technical writing is an incremental—and iterative— process. This section guides you each step of the way, from sitting down in your seat at a new job, to researching, to writing drafts, to handing in your final product, and finally to patting yourself on the back for a job well done.

.

Chapter 13

Getting Started

Get your tech writing feet wet by walking into your first job and being productive from the start.

What's in this chapter

- Starting on the path to success
- What your first assignment may be like
- Document maintenance
- Why now is a good time to ask for help

You may be surprised to see a chapter called "Getting Started" so late in the book.

But up to now, we've been building your foundation, the foundation that helps you to understand all the parts of what make a good technical writer and good technical writing. As you get started in your career as a technical writer, you can use the information in the rest of this book to succeed on the job.

This chapter will guide you through getting started on the job, as you walk into a new environment and figure out what to do. There's also a lot of information here that will help you avoid common pitfalls and wrong turns. If your ultimate destination is to be a consultant, you'll find this chapter especially useful since you will be walking into a new environment many times.

So think of your journey of a thousand tech writing miles as beginning with the first step, and step right in.

New Kid on the Block

Although much of this book talks about writing documentation from scratch, your first foray into technical writing probably won't take you down that long and winding road. When you start a new job, you're most likely to be given proofreading or editing work on something that already exists. While you do that, you may also be reading existing documentation about the company's products so you can learn about them. Soon, you'll take on a project that requires updating.

Taking over someone else's work requires tact. It's bad form to complain about how badly something is written and brag about how you've improved it. You don't know the previous writer's circumstances—because of limited time and resources, it actually might have been considered a glowing success. You could even discover—too late—that the writer you are criticizing is your manager! So if you can't say anything nice...

It's always easier to revise and improve what already exists than to create original content from scratch. And you can apply your writing skills while you learn about the style and structure of the company's documentation and the nuts and bolts of the company's technology.

Before long, you could even be responsible for all documentation for a product family that will require content from scratch. Or you might work, along with other members of the Tech Pubs team, on a number of documentation deliverables across many product families. In the high-tech world, those chances could come sooner than you think. A tech writer's world is nothing if not fast-paced!

All managers have their own ideas about how much time it should take before you start contributing. Some want you to jump in and be productive on Day One and some will give you several weeks to ramp up while gradually doing productive work. This is a good thing to find out about at the interview—and be prepared to behave accordingly.

Ask About the Style Guide

When you start at a new job (or even before you start), ask your colleagues if there is a style guide to consult. Style guides, which are discussed in detail in Chapter 18, "The Always-in-Style Guide," should tell you how to spell product names, what terminology to use ("Web" with a capital W or lowercase? "Log in" or "Sign in"?), and what typographical conventions to use.

If the company has a style guide, read it, learn it, and apply its rules to the material you are asked to work on. If there's no style guide, review similar documentation so you have a sense of how things are already done. When you come across something questionable or inconsistent, refer to the existing material for a guideline, and then make a note so you can ask your manager about it later.

In fact, keep notes on all the style issues you see. The notes you take can become the basis for a future style guide.

The Care and Feeding of Your First Project

If you've never worked as a technical writer before, you may wonder what you're expected to do when your manager announces you are responsible for "updating" or "taking over" existing documentation. What exactly does that mean?

Congratulations. You've just adopted a project—some kind of document, help library, or other content. It's not unlike bringing a pet home from the animal shelter. No matter what its history or how it came to be the way it is, its welfare now depends on you.

Ask your manager, or the person responsible for the project, what his expectations are. Make sure you are clear on what your responsibilities are, and don't be afraid to ask if you don't understand. Are you proofreading? Doing a major rewrite? Adding updates for a new software release?

As you begin to work on your project, make sure that you add new information appropriately, along with fixing known mistakes or omissions. At the same time, look for style inconsistencies or opportunities for improvement. If a sentence or paragraph is unclear to you, it's likely to be unclear to everyone else, too, so mark it and plan to investigate.

Ownership of a document can be pretty fluid in the high-tech world. You could be asked to update sections of a deliverable while another writer is considered to be ultimately responsible for it. Or you could be asked to work on something for some weeks and then give it back to its original owner—just as you started to feel it was yours. There is often way more work to do than hands to do it, so it's not unusual for a Tech Pubs department to pass work on from one person to another as priorities shift, or for more than one person to work on a single project. Don't be surprised by any of this and don't take it personally if you labor over something only to have it given back to a different writer.

A document can change so much over different revisions or software release cycles that one day you may not even recognize your contribution. And that's as it should be. After all, the documentation is about the product, not about you or what you did to it.

How Much Rewriting Should You Do?

Although the wordsmith in you might be itching to revise or completely reorganize parts of the project, be cautious about how much of this you do. In today's work world, there's not always a lot of time for excessive polishing. You must balance time spent rewriting against time making substantive, necessary changes, and make sure your manager is on board with both.

If a document's information is accurate and complete, rewriting and reorganizing may be unwise even if you think you could tremendously improve it. A user accustomed to seeing information organized in a particular way may not appreciate having to relearn everything in the next revision of the documentation.

Maintenance and Tune-Ups

Not all tech writing will be a daring adventure, scaling the peaks of new technology or channeling your internal Hemingway. Some of it is plain old plodding, just putting one foot in front of the other.

Users depend on the documentation to be accurate and to reflect the current state of your company's product or service. Products are always being upgraded, improved, and changed. In the case of software, that can mean many iterative releases and patches. For hardware, it can mean anything from a minor tweak to a completely different industrial design. So it's crucial that documentation be regularly brought up to date. Maintenance, while not as exciting perhaps as writing brand-new material, is critical to customer satisfaction.

If it ain't broke, don't fix it. Wading into a major rewrite and reorganization simply because you think your ideas are better is a habit of technical writers everywhere, but it's not always a good idea.

Consider how much time your rework will take before you start. It might be something your manager thinks is a waste of time. Discuss proposed big overhauls with your manager before you start and find out a few facts. Are customers satisfied with this documentation as it is now? Is the content reused in many places so your rework might break some other structure? Is this particular product being phased out or does it have a very small customer base? It won't make you look good to spend valuable time on a job that didn't need to be done in the first place.

Knowing What Needs Fixing

If you are asked to do this kind of maintenance, that means your first priorities will be to add new information to address the new release, make corrections to known errors, and fill in the missing pieces, whatever they might be.

Changes and corrections aren't always collected for you in a tidy list. You'll find them in a variety of places, and many you'll have to ferret out for yourself, by attending developer or scrum meetings and talking with engineers, product managers, designers, and customer support personnel.

Not every suggestion you receive needs to appear in the documentation; as you become more knowledgeable, you'll use your judgment about which ones to include.

Working on revisions can take nearly as long as doing the original work. You may think a document needs a little change here and a little addition there, but even a routine update may need to address all of the following:

Don't revise and rewrite a document just for the sake of improving it, unless you are also adding new content to describe new features and functionality. A user assumes the content is new if it's presented differently and it can be annoying to read and review what appears to be new material only to discover that it's the same old stuff.

- Update document ID numbers, release or version numbers, and dates.
- Describe new features and functionality added to this release. In the release notes, describe the bug fixes that were made.
- Correct known errors.
- Add information that is missing.
- Clarify and rewrite as necessary.
- Update the index.
- Make sure all screen shots, diagrams, tables, and other illustrative matter are up to date. Reviewers tend to overlook them when they read drafts.
- Update the document's template.
- Find out if any customers or partners had change requests or feedback and address them.
- Update and test cross-references and other links.

A Plan of Attack

By now you may have an idea of where you want to go in updating a document, but might not be sure exactly what to do first. Here are some guidelines to point you in the right direction.

- When you receive your first assignment, read it, or at least skim it, as soon as possible. Your manager and the subject matter experts you meet with will expect you to be familiar with the document's contents. You should be able to answer questions about what is and isn't in the document and have some thoughts about what changes should be made.

- Take a look at the product or service and use it yourself, or ask for a demo from one of the developers. If it's not yet in the prototype stage, ask where you can find specs or design documents that will help you understand what the product is about. Refer to Chapter 10, "Become Your Own Subject Matter Expert," for ideas on how to learn about the technology.
- Find out who are the go-to people for your subject and seek them out. Go introduce yourself and tell them what you're working on and that you'll set up a review meeting. Ask if they have anything for you to read before the meeting, and then be sure to read it. Read Chapter 14, "Gathering Information," to learn more about working with all of the people who are your sources of help and information.
- If you are taking over a project from another writer and that writer is still available, talk to her about the content and the changes needed. She might have been unable to do everything she wanted to do on the project and can tell you what some of the important issues are.
- Look at comparable documentation for other products and see how it was made and what its contents are.
- Turn on change bars to indicate every place where you have made an update. When you think you have all of your changes in place, ask for a face-to-face meeting with your manager or the writer who is overseeing your assignment. Sit with him and go over the work you have done. It will be a lot easier for you to understand what is right and what can use improvement if you talk to him rather than sending a draft out through email.

When you're ready to send the documentation out for review, make sure you follow the review processes within your company. See Chapter 16, "Everybody's a Critic—Reviews and Reviewers," to learn more about reviews.

Making Your Deadlines

It's not unusual for your first assignment to come during a real crunch and for you to be asked to do something faster than you ever thought possible. After all, you were hired because there was a shortage of tech writers, and your manager may have been saving up this work for a while.

You might find yourself in a situation where there's no time to learn about the product and your boss expects you to do a job that, from where you start, looks like jumping the Grand Canyon without even being allowed a running start.

Estimating your time and meeting deadlines can often be a trouble spot for a new tech writer. (Hey, it's difficult for a lot of old-timers, too.) At the beginning of an assignment, you probably aren't even thinking about the deadline, although it might be a mere few days—or hours—away.

Check Your Progress

Keep a close watch on your progress. If you feel you might not be able to meet your deadline, let your manager know the moment you feel you are having trouble. It is never OK to miss a deadline you've agreed to meet, but it *is* OK to realize well in advance that circumstances have changed and you are not sure if you'll be able to accomplish everything that's needed.

When you go to meetings for the first time for a project with which you're unfamiliar, don't feel bad if you don't understand anything. The acronyms, corporate buzzwords, and technical references will sound like a foreign language.

Just sit and listen. No one will expect you to contribute anything, and if they do, don't pretend to know what you don't—just tell them you are new and still getting up to speed. One day, all will suddenly fall into place, and you'll be fluent in the language that once seemed so foreign.

This gives your manager time to resolve the problem. There is nothing more frustrating for a manager than to learn that someone is missing a deadline—when it's too late to do anything about it. By breaking your schedule into mini milestones right up front and following the guidelines in Chapter 11, "You Want it *When?*"you'll be able to manage your time effectively.

No Matter What Happens, Stay Calm

Many newcomers are shocked by the pace in high-tech companies. It may seem that your early assignments are impossible, and require super-human abilities to finish. Slow down. Take a deep breath. Then map out your plan. Stay calm and focused—and always aware of your deadline.

As a professional, it's important to work like a duck: look cool and calm on the surface even when you're paddling like crazy underneath. But please, don't be so cool that you appear to be arrogant, or act as if you know everything already.

Finishing the First Assignment

So congratulations, you worked like crazy, and had to stay until 10:00 PM to do your final formatting, but you finished your first assignment at the new job and it looked awesome. When you stumble bleary-eyed into the office the next morning, you wait eagerly for praise from your boss, a pat on the back for accomplishing what had seemed the impossible.

And instead of the "kudos" you expected, you hear, "I found a few errors in that PDF you generated. I marked it up—can you fix them before lunch and post a revision?" You may feel angry that she checked on your work, you may be surprised that she was up working even later than you, and you may feel completely discouraged that no matter how hard and late you worked, no one seems to appreciate your effort.

Don't take it personally if you don't get enough praise for completing your job on time. Tech Pubs managers in today's fast-paced working world are under huge pressure to deliver. They don't always remember to give praise for what to them is just doing your job. But it doesn't mean they don't appreciate you.

Once the job is done, ask your manager if you met the expectations and what you can do better next time. Discuss the parts of the assignment that were difficult and the parts you think went well. This will give you a good chance to find out how you did and what your manager thinks. By approaching the situation in a businesslike manner, you can avoid hurt feelings and also improve your work.

After you've been on the job a while, you may come to realize that the first assignment you thought was so scary wasn't such a big deal at all.

Ask for Help

Above all, realize that as a new hire, it's the last time you'll be able to ask all the dumb questions you want, and no one will criticize you for your lack of knowledge. Yes, everyone is busy, but most people are happy to assist if you are honest and specific about what you need.

Take advantage! You have nothing to gain by pretending to know more than you really do. And as you tread the path from newbie to old-timer, you'll soon be the one helping the latest writer on board.

Chapter 14

Gathering Information

Knowing what to look for puts you halfway there.

What's in this chapter

- Scavenging for information
- When to read and when to listen
- How to work with developers

We've learned a lot about how to find out as much as you can about the product, by reading existing documentation, learning how the customer uses it, and trying it out yourself. All of that is indeed important. What you'll find, though, is that often your most important source of information is your subject matter experts—the group of developers and product managers and QA and Customer Support people with whom you interact every day.

Building a good relationship with these people is critical. You'll depend on them to let you know about changes to the technology. You'll depend on them to review your work. You'll depend on them to let you know of changes and additions that should be made to your documentation. And until you learn the product really well, you'll depend on them to explain how things work.

It's that last part that can be a bit tricky for a technical writer. How much of your job is like that of a reporter versus that of an information developer? Whose responsibility is it to do detailed reviews? In an industry where everyone is busy, is it OK to ask an engineer to write something for you, so you can just edit

A small digital recorder is an invaluable tool for you as you talk to different experts in the company. It enables you to later replay everything that was said, including those parts you missed. Best, the tape recorder frees you from taking such detailed notes that you miss the gist of the conversation. Instead, use your laptop or a good old-fashioned pen to note the highlights of the information session, and use your saved energy to ask the right questions.

it? This chapter helps you answer some of these questions as you go through the ups and downs of gathering information from your technical sources.

Scavenger Hunt

Collecting the information you need to develop good documentation is like a scavenger hunt. You start with some clues about the information you need, and as you gather that information, one fact at a time, you make progress toward your goal.

Keep all of the information related to a single project in a single place, and make sure you keep everything. If you use a ball-point pen and notebook or digital recorder, carry it with you whenever you go on the hunt. If you store all of your information in a certain folder on your laptop, make sure you are rigorous in keeping those notes and keeping them in the same location. Back your material up on a thumb drive or store it in a version control system to make sure it is safe.

As you work on the project, go through your notes regularly. Some things that made no sense to you early in the project will make sense later as you gain a greater understanding. Check off items as you incorporate them into the documentation.

Read, Read, Read

As I've already mentioned, your starting point as you gather information is to read everything you can that pertains to your project. If you are lucky, there will be requirements documents and design documents available. If the product is mature, there will be older versions of the product documentation as well.

Start out by reviewing the requirements documents, which should neatly condense the purpose of the product as well as the features that enable it to meet the purpose. Don't assume that everything in the requirements document has made its way into the final product, though. Often, as development progresses, the team discovers they don't have time to do everything that's asked for. The requirements are re-prioritized and some features and functionality are put on hold. And the product manager doesn't always go back to update the documentation to reflect reality.

Listen, Listen, Listen

Chapter 10, "Become Your Own Subject Matter Expert,"talked about all the ways in which you can, and will, learn about the product as you live and work with it. But no matter how much you read, or work with a prototype or the actual product, at some point you're going to need to go to the people who drive and develop the product itself. These subject matter experts (SMEs, in technical-writer-speak), can often be your best source of information. And how do you get that information from them? The old-fashioned way—by talking to them. By interviewing them.

> **INSIDERS KNOW**
>
> You might find you need a lot of persistence in chasing down the elusive SME. Some people are just so busy that no matter how many email messages you send or how many times you show up in their cube or office, they just don't respond to you.
>
> After a while, it's a drain on your time and your energy to continue to chase the same person. Try to get the information you need from someone else. If that's not possible, as a last resort, ask your manager to intervene.

Relax, nobody expects you to be Oprah Winfrey. An interview is simply two or more people getting together to talk about something they have a common interest in. One person (you) has a set of questions for which you would like answers. The other person, hopefully, has the answers, and has the time and willingness to give them to you.

"Interview" is too formal a term to use when you're setting up a meeting, but do be specific about what you want to talk about and how long you want to meet. If you need a product demonstration, set up the meeting for a half hour, or hour, or whatever time slot is acceptable at your company. If you only need 15 minutes to ask questions about a specific function, say so. If you can, send some of your questions in advance, so the developer can start thinking about them.

Be respectful of your co-worker's time. By the time you meet with her, you should have a fairly good understanding of the topic you're writing about. Although it's certain you won't know everything, make it your responsibility to know enough so you can ask intelligent questions.

But there's no need to pretend to understand something when you don't. Be very clear about what you don't know and listen to what your co-worker is telling you. Avoid the temptation to interrupt. Remember, at this time, you are the one who wants answers.

Before the meeting ends, review your notes and make sure you have answers to your questions and you have cleared up any points about which you are unsure. You also may want to ask if there's anything the other person can think of that you should know but that you didn't know enough to ask about. This could provide you with an essential piece of knowledge you may never have discovered otherwise.

Keeping Up

If you are watching a developer or expert user take you through a product step by step, you are at a disadvantage. Because you know less than the person doing the demo, you must not only watch something new but also ask questions and take notes. You'll find yourself repeatedly asking (or wanting to ask) the demonstrator to go back and show you something again or tell you how something was done. Don't get tense or apologetic about this; you're doing your job. The demonstrator's job at that moment is to help you.

Human beings are by nature highly visual, and software engineers seem to be more so than most people. Many developers you'll work with think in images rather than words. Bear this in mind when an engineer starts drawing—instead of talking—to explain a point.

It's a good idea to have a camera or notebook with you so you can accurately copy the drawing. A flowchart or illustration can be an excellent way to explain something to your end user.

Do make an effort to keep things moving. And don't wait until the end to say thank you. Let your colleague know how you appreciate their effort. Like you, developers are very busy and however gracious they are, they don't always like to interrupt their work to talk to a technical writer. Some developers resent the process, feeling that the tech writers expect too much from developers—demos, reviews, answering questions, even at times writing the documentation.

Write First, Ask Questions Later

I am a firm believer in writing first so you can ask the right questions later. Whether your assignment is to do minor updates or write a brand-new piece, the best way to get information from other people is to first try to figure out everything yourself. Instead of going to a developer with the expectation that he'll tell you "everything," by the time you go to him, you should have a manageable set of very specific questions.

As you read, research, and assemble the information you've collected, start to write. Do your best to develop all your information with the knowledge you've got. You will find as you write that you make statements whose accuracy you are unsure of. Make a note so you can ask someone. You will also find that there are areas you think need to be fleshed out, but you have no idea what they are. Make another note.

It's important to keep the questions specific. If you don't, it means you don't understand the content, and when you don't understand the content, you're not likely to know the right questions to ask or understand the answers you receive.

Avoid a situation where you ask, or expect, the developers to write content for you. That is your job, not the developer's. (If a colleague does write something for you in the interest of saving time or because it's easier for him to write it than to explain to you, be aware that this is a tremendous favor, not business as usual.)

> **INSIDERS KNOW**
>
> We all get a *lot* of email during the work day and some of it is lost or ignored. If you're dealing with a reviewer who doesn't seem to respond to your email, try meeting with him face to face.
>
> Take a paper copy of what you're working on to his desk and ask to sit with him to go over some questions. Or print out only a short segment of your documentation so he isn't intimidated by the thought of having to review a huge manual. For many people, the face-to-face contact is far better and faster. You just have to figure out who those people are.

After a session with a developer, go back to your desk and immediately add what you've learned to your outline or draft. If you don't do it right away while it's fresh in your mind, one day you'll pick up your notes and not understand a word.

Don't be one of those annoying tech writers who comes back again and again to ask the same questions. This happens if you don't understand the material and you ask for too much information in an attempt to learn it quickly. Ask for only as much as you can absorb, and tell the developer you have to stop and work on the writing when you reach your limit.

Live and In Person

I suggest that if possible, you avoid using email to ask questions of your subject matter experts. Sending a broad question—one that requires a lot of work to answer—in email can be very off-putting and again, you are asking the developer to do your work for you (writing). If you do need to use email to ask questions, keep the request to one or at most two questions that require very short and specific answers, and send the message to one person only.

Email blasts with a bunch of questions tend to go unanswered, since everyone thinks someone else will answer it. Remember that not everyone likes, or has time, to write long email responses to questions.

Eyes on the Prize

No discussion of information gathering would be complete without mentioning the value—the necessity—of both patience and persistence. You'll need them both, and often.

This work involves dealing with a wide range of personality types, and they all know something you need to slice, dice, and absorb. Technical writing is a continuous process of learning, carefully gathering, sifting, organizing, and assessing, all while trying to craft something that makes sense for a user. You might find that you ask one person a question and that person points you to another one, who then points you to another. You may discover that you receive three conflicting pieces of information from these subject matter experts, and that you have to start over. It's not always easy!

If you have a limited number of specific questions for a developer, ask these questions face to face, or (if the developer works at another location) over the phone, until you develop a productive email relationship.

But neither is it a constant battle with the forces of chaos, evasion, and delay. Most people honestly want to help you deliver high-quality documentation and they will do their best. Do your best by making sure that you make the most of their limited time and you come to them fully armed with knowledge. The focus you bring—in the form of patience and persistence—is one of your most valuable assets.

Chapter 15

Putting It All Together

Time to start writing.

What's in this chapter

- No more planning—let the writing begin
- How to turn an outline into a finished work
- Some ideas for breaking writer's block

After having read that last 14 chapters, you should have a good idea of how to accumulate everything you need to start creating some content. You've found a job, started your first day, learned all you can about the subjects you're writing about, figured out what some of your formatting options are, and created a documentation plan. It's finally time to start putting fingertips to keyboard.

This chapter will help take you through some of the steps involved in creating an entire deliverable, or set of deliverables, from scratch. You'll write an outline and a first draft. You'll find that filling in an outline is a lot easier than simply sitting at your computer, banging away on your keyboard, and hoping something coherent comes out of it all.

You'll also learn some tips to help you get over writer's block, or just the procrastination that sometimes happens when you're faced with a huge pile of input and you're not quite sure what the output should be.

Following the Rules

Your department is likely to use a template as a basic structure for documentation, and this should be of some help as you develop your content. A *template* is a file that contains the layout, design, and sometimes the basic elements that belong in the documentation. (Chapter 20, "Design and Layout," talks more about templates.)

Even if you don't use structured content, your department is likely to have requirements about what sections go in certain deliverables. For example, a release note might require sections called "About this Release," "Product Compatibility," "Prerequisites," "New Features," "Fixed Issues," "Known Issues," and "Contact Customer Support." When sections are required in specific order, it saves you some time in figuring out what goes first as you put together your content outline.

If your department doesn't have guidelines to determine what content and structure belong in different document deliverables, consider creating some. It is helpful to writers starting on a new project and it provides consistency for customers who search for information.

Jump Right In, The Water's Fine

By now you should know a lot about what type of documentation you want to produce and whom you are producing it for. You should have some knowledge of your subject matter and have collected a lot of raw material, in the form of notes and existing content, to contribute to your documentation output. You're ready to write...even though it might feel a little scary.

The nice thing about writing is that you don't need to write in a linear fashion. Whether you're writing a manual, help, or a data sheet, you can write in whatever order you like and assemble it all later. As long as the parts fit together correctly in the end and the content is complete, it doesn't matter in what order you wrote each piece. So start typing!

And this is where an outline helps enormously. If you've prepared an outline, even a sketchy one, you can easily figure out what goes where in the end.

In the Beginning: Creating an Outline

All of the bits and pieces of information you've collected to date are like pieces of a jigsaw puzzle. When you dump the whole box onto a table, you know the whole picture is there—you just have to organize the pieces to see it.

Creating an outline for your document is a lot like starting on a jigsaw puzzle: different people start in different ways. Some connect all the edge pieces first; others group the pieces by identifiable colors and patterns and build sections.

A book typically has a beginning and end with some kind of progression in the middle. Online documentation, including help, has a tree of information that connects in multiple ways and on many different levels. Online information must be tracked carefully to make sure the user can navigate easily from one place to another. Even with these differences, you can use an outline to organize your information and put it in a hierarchy.

Outline Format

Classic outlines use capital letters to indicate the top-level headings, then numbers, and then lowercase letters. You can continue to build the outline by alternating letters and numbers, indenting each level. If you use Microsoft Word, you can use its built-in outlining feature.

The classic outline format is fine, as are many other methods you may choose, as long as it works for you. You might just jot down major headings and subheadings, or you might be a super outliner who continues to fill in topics and information until you've written the entire document! Either way is fine.

Start your outline by putting in all the required sections; that is, the sections that are required according to your department's template or schema. If you don't have requirements, put in section headings you see in similar documents or help systems—you can always delete them later if they aren't right.

Your first pass at an outline might look something like this:

EnterPrize Cloud Storage Installation Guide

A. Introduction

1. About EnterPrize Cloud Storage
2. About This Guide
3. Installation Process Overview

B. System Requirements and Prerequisites

1. Hardware Requirements
2. Software Requirements
3. Security Requirements
4. Before You Start
    a. Network configuration
    b. Server setup

As you fill it in, you'll add more subtopics under the headings, and add more headings as you think of them. At the lowest level heading, drop in bullet lists of topics and ideas that you think belong in the section.

The outline can be used for shared content. As you decide what goes into the user guide and what goes into the help, use color-coding or strikethrough or other cues to indicate where the content belongs. If "About this Manual" and "Troubleshooting" go only in a user guide, make those lines a color that shows that. If some sections go only in the help, give those entries a different color

Shuffling the Virtual Index Cards

Writers used to fill in their outlines by writing all their topics on index cards and shuffling them around in piles, laying them where they seem to fit and moving them if they weren't right. It was kind of like the single-sourcing of yesterday. Luckily, we now have word processing!

The method is the same whether you use index cards or a white board or a computer, though. Your goal is to collect all your notes, email messages, content fragments, and existing material, and place each piece where it best fits. Start plugging all the content you've gathered into your outline. If you took notes with a pen or pencil, type them up and put the content wherever you think it belongs.

Use parts from documents that already exist, if you think they fit, and copy them right in. (It's not plagiarism to incorporate existing material from documents your company owns. It can save you a lot of time and effort if such material is available.) Graphics, text, and whole blocks of information from marketing materials, specs, design documents, and company wikis might be just the right thing to fill in the holes.

Make sure the material is still up to date, though! Using obsolete information isn't going to help anyone, and trying to unweave it all later can be a real mess.

If you are unsure how to start, start out by breaking your content into the smallest complete chunks, and assign each chunk to one of the following categories (you may recognize them as DITA topic categories):

- **Task** for a procedure that describes how to accomplish a task
- **Concept** for about a product, a task, a process, or any other conceptual or descriptive information
- **Reference** for topics that describe reference information in support of a task

Group your content into collections of related pieces and processes. Don't worry about order at this point; focus on relatedness.

Start at both ends. What content chunk is introductory or foundational? Put it first. What chunks are supplementary? Put them later. If a content chunk is interesting but doesn't seem to go with the flow, maybe it belongs in an appendix. What's the first thing a user needs to do? Or what's the most common thing a user needs to do?

If you find some content that just does not seem to belong in your outline, it doesn't always mean that you need a new heading in your outline. Instead, maybe this chunk doesn't belong in this particular deliverable. Set it aside for now; there's every chance you will want to use it later.

When writing narrative (book-style) documentation, generate a table of contents early and often so you can check on the structure of your chapters and of the book as a whole. You'll notice when sections are in the wrong place and you'll notice gaps more easily than if you were reading the entire thing.

Go through every single piece of information you have gathered and methodically deal with each piece—either by adding it to an appropriate level in the outline, or by adding it to another file you create for later use, or for use in another document. Don't delete anything yet—you don't know at this point whether something is really relevant.

By the time you are done with this, you will have placed every single piece of content you've collected into a structure. Congratulations! You've got a document or a help set. It's not yet beautiful or polished or woven together, and it's probably got a lot of gaps, but you should be able to see an identifiable whole.

Help Is Not Linear But Sequence Matters Nonetheless

When developing help, you need to predict everything a user might want to see. Beginners need step-by-step procedures. Users with more expertise want to have their questions answered. If a help topic can lead to more detail, it's a good idea to add some links to the related topics.

You also need to think about whether the user has actually landed on the right topic—maybe you need to suggest a list of the topics they really want to see rather than the one they're looking at.

And you, the creator, need to think in terms of a big network of topics that interconnect in many ways. Remember the topic-oriented writing discussed in Chapter 8, "The Deliverables." The user does not need, or want, a sequence of events that starts on page 1 and continues logically through to an end point.

Instead, the user wants information relating to a specific subject—the subject matter in use at that moment. However, each topic is a building block that can lead naturally to one of many other topics. Users make their way through this maze by way of searching, using a table of contents or index of topics, and by links within the help topics themselves that lead to related topics.

If you are using shared content, make sure your content chunks are broken into the smallest relevant piece for an online help user. Let's say that the most important material to provide in help is procedural information. Maybe you decide that you will provide no conceptual information in help and save all of that for a user guide. Or maybe you decide that you will provide conceptual information, but it stands alone as its own topic and will be available only when a user clicks a link to ask for it.

Here are some guidelines to help you get started writing help:

- **Make each topic short and relevant.** Users don't want to scroll through a massive tome when they need help. Eliminate all but the essential information. If a user might need background information or more detailed information, put a link at the bottom of the topic with related information.
- **Tell the user what about related information.** Don't be vague here—you're sending your users on a chase through the information maze, so tell them exactly what they're getting into. Instead of a link saying "more information," provide specific information in these links, such as "Removing a user," "Changing user privileges," and "Didn't want to add a user? See *Adding Accounts.*"

Planning—Help!

It can be helpful to start out by mapping your help system with a tool such as Microsoft Visio. This way you see each main topic and can draw links to related subtopics.

Let's plan help for our company's new consumer application, HomePrize Cloud Storage. Start simple. During an initial planning session, you don't have to think of every single topic that can possibly exist, and you don't have to break the information down into minor subtopics yet. That will come later as you develop the content and, if possible, test it on potential users.

Consider your audience as you create a high-level list of what users need to know. This is a consumer product, so getting started must be easy. So let's have a topic about that. The Getting Started portion should enable a customer to start successfully using the product without dealing with all the advanced features and functionality of the product. A Getting Started section might contain a video tutorial as well as text topics.

And then there are the different types of users. There will be users who have no idea what "backup" and "syncing" are all about, along with users who fully understand these terms. How do you meet the needs of both of them? Maybe your inexperienced user needs some simple conceptual information in addition to the "how-tos" of getting started.

Next, start thinking about all the features and functionality. What would a user expect to do, at a minimum? Does each need its own topic? Is there a natural way to group them together? Is there some type of linear progression required in any set of procedures?

You will also want to answer questions about the customer's account—how to renew, how to upgrade, how to check account status. And how to set up the application on different types of computers and mobile devices. And there's troubleshooting. What happens if a user loses—or thinks he's lost—his data? What if files won't upload? What if he runs out of space?

As you start thinking about all of the possible things a user needs to know, you'll see that you can end up with a very large number of topics. Keep track, in outline or diagram form.

Drilling Down

As you flesh out the content, start developing the ways in which the topics will relate to each other—as shown in the outline, what is a high-level topic, what is a subtopic, and what topics should link to other topics? The system is like a tree, with each limb growing branches that in turn produce their own branches. The user has the ability to jump around from one branch to another, but only if you provide the means. This can be done by always going back to the central location—the tree's trunk—and moving to the appropriate branch, or by jumping straight to another branch that you have decided contains relevant information.

Don't let any topic be a dead end. It should always link back to something else, even if it's just the *splash page*, that is, the "title" page of the help.

If you don't develop context-sensitive or page-sensitive help, a splash page or the first page of your content is displayed every time the user clicks **Help**. The splash page should include the name of the product, your company's logo and information, and links to key areas of the help.

Fleshing Out the Outline

It's time to start gluing all of your content chunks together, with transitional information and by filling in missing technical content that belongs in there. Choose the chapter or topic that feels most complete or that you feel most confident about, and start writing. If there's no clear favorite, just start at the obvious place: the beginning.

Always start with the broad view and continue by funneling down into more and more detail. Making a declarative statement about the topic at hand is a good way to get momentum going—"EnterPrize Cloud Storage 2.0 is ..." or "Before you install EnterPrize Cloud Storage 2.0, you must ..."

And always give users a context. Tell them how they got to where they are. Tell them the purpose of what they are about to read or do, what to do first, and why they are doing it. (The what and why might not always be present in output such as online help, although it is a good idea to provide that supplemental information by linking to it.)

Drafting Your Draft

The first few paragraphs or topics you write could very well end up being scrapped. And that's OK. This is a draft, not a finished piece, so don't worry now about whether the wording is perfect or even if the organization is perfect. (Do, however, be worried about the accuracy of your information.)

Writing is an iterative process, and it is not always going to be perfect the first time out. Sometimes you have to start getting the words down to even know what the result might look like. The important thing is to write. You can always rewrite, revise, or reject what has been written, but you can't do any of that if you haven't written it in the first place. You can't correct what doesn't exist.

Author Comments

Expect to have a lot of gaps at first, and a lot of questions you can't answer. As you read what you put together, put yourself in the shoes of the customers and think about what questions they might have. If you can't answer these questions yourself, embed them in the text as notes to yourself or to your reviewers, like this:

Remove unneeded files. ***[Reviewers: Which files do they remove?]***

Make these questions specific. If your author comments are too broad and you're asking for too much information, no one will answer them.

Your author comments should also be used as placeholders to indicate that you intend to add some information later. You can add a single line saying "TBS," or you can say more, such as "I'll include screen shots for the File Manager application as soon as the UI is completed." This way everyone can see what's missing and that you know about it. Instead of stopping to think about what's not there, the people who read your draft need to know that you will take care of it.

Leave Your Mark

Make absolutely sure that your draft contains unmistakable cues that this is a draft—the word "DRAFT" in the header or footer, or as a watermark on every page is a good clue. The file name should also have the word "draft" in it: "EnterPrize_Cloud_Storage_2.0_Install_DRAFT2.pdf" is a name that includes the product and version, the type of manual, and the fact that this is a second draft. Anyone can easily understand from the PDF name exactly what kind of document it is and what phase it is in.

Even though a first draft is usually going to be very incomplete, it's still a good idea to proofread and spell-check your work before sending it out. A sloppy draft will turn off your reviewers and make them feel—maybe subconsciously, maybe not—that your content is incorrect.

It's surprising how draft copies seem to pop up in the most unexpected places: a sales rep or product manager will give a copy to a customer, or a network administrator will try to work from one of them, or a customer support rep will refer to a copy when on a customer call. If they don't see the word "Draft" on the page or in the file name, you can't blame them for thinking this is the final word.

If a customer calls tech support complaining about incorrect steps on page 21 of *EnterPrize Cloud Storage 2.0 Installation Guide* and they don't match the support rep's steps on page 21, the problem gets worse. The support rep looks bad, the company looks bad, and the Tech Pubs team looks very bad.

Besides marking a document as a draft, make sure you follow company guidelines to indicate whether a document is proprietary. If your final document is to be confidential and proprietary, make sure you also mark every draft as confidential and proprietary.

Between the information chunks you've dropped into your outline form, your transitional writing, the gaps you've filled in based on your own knowledge and research, and the author comments you've added to the draft, your draft should now be very complete.

Embed your comments, questions, and "To Be Supplied" (TBS) markers into your draft so the reviewers see them, but make sure you set them apart in a way that ensures they won't make it into the final product. (It's embarrassing when that happens!)

You can avoid accidentally leaving the comments in by creating a style for author comments that puts them in a different color, bold, all caps, or even all of the above. Use it faithfully and turn it off to hide all comments before your project is released.

If It's Incomplete, It's Still a Draft

A first draft is your first pass at assembling information into its final form. It should be at least 70 percent complete, with content fairly well organized and accurate, although some content will probably be missing. Writing should be acceptable but does not have to be polished. Placeholders should clearly indicate where graphics such as screen shots, diagrams, and tables belong.

The second draft incorporates reviewer feedback and is often the final draft. With the second draft, aim for at least 95 percent completion. Content should be accurate and well-organized. The writing should be good with smooth transitions. If there are any areas about which you have any questions, you must clearly mark these for reviewers. Most, if not all, of the graphics should be in place.

How Fast Is On Time?

While juggling your flaming sticks, you also have to do a balancing act—balancing the time you have available to write a draft with the amount of content you're able to get into that draft. Keep in mind that the function of a first draft is not necessarily to dot all the i's and cross all the t's, but to get the content in place with as much structure and accuracy as possible, to get confirmation on the document's basic outline, and to elicit answers to questions. Writing a first draft is like putting the first coat of paint on a wall—get it on, knowing you'll apply the finishing touches later.

Your documentation plan should have your draft due date in it, published for all to see. Do whatever it takes to meet this schedule. If you're not going to be able to meet it through no fault of your own, make an early decision about whether you want to try to push out the review date, or if you are going to send out the draft as is on the date you've already scheduled.

Polishing the Rough Edges

No matter how rushed you are, do leave some time to make your draft look as good, and finished, as it can be before you send it out. Read it to make sure it flows and the language is clear. Run a spell-check. Fill in all the holes and gaps, if not with information, at least with "TBS" or author's questions.

Why? Because a sloppy draft can be a distraction for your reviewers. The person reading it will unconsciously think you don't know what you're doing and they will either not pay enough attention to what *is* correct, or they will be distracted by errors and start correcting your language and spelling. It's unfair of you to put the burden on your reviewers to correct your mistakes.

Battling Writer's Block

When deadlines are looming and the words just don't flow, try some of these tips for getting in the writing groove:

- **Block out some time**. And I do mean block. Set your calendar to busy, turn off the phone, close the door (if you have one), and give yourself time to work. If you can't work at your desk without being constantly interrupted, book a conference room for a few hours. Or work from home if home is a no-interrupt zone. I'm talking about time in increments of hours, not minutes, as many as you can set aside. It takes some time to gather serious writing momentum. Writers often think they can squeeze in some work during a quick break, but it really takes more than half an hour to gather serious writing momentum. (If you have only a short amount of time, work on email or indexing or other tasks that can be started and stopped. Writing takes more time.)

When you're in a hurry, you'll be grateful for the accommodating engineer who writes some content for you to drop into your documentation. Even if the material seems ready to go, make sure you understand it before dropping it in place. The part you don't understand is the part someone will later ask you to explain. Remember, if it's not clear to you, it won't be clear to someone else.

- **Just do it.** Pick a time, set an alarm, and get moving when the alarm goes off. Don't wander off to get another cup of coffee, don't take another peek at your Facebook page, don't chat with co-workers. Just do it.

  "It's not that easy!" I can hear you say. But it is. It takes discipline to write. And yes, staring at an empty FrameMaker template or blank screen is daunting. But waiting for the muse to strike is not an option. Just put your fingers to the keyboard and go.

Use your spell-checker! Don't provide some unintended humor by letting some careless errors slip by.

And don't let the spell-checker make you a victim of the "Cupertino effect," in which an application's spell-checker or auto-correction replaces a correct word (not in its dictionary) with an inappropriate word. *Cupertino* (Apple's home town), was in many programs' dictionaries when other words were not. Applications kept changing the word "cooperation" to "Cupertino."

- **Start anywhere, with anything.** Sequential thinking is a great ally when you're working through a process or outline, but there are times when it's not your friend. If you think you should begin at the beginning but find yourself doing nothing because you don't know where to start, start anywhere. If you don't know what to write, retype something from an older version of the documentation. Or list five things you think describe the product. Sometimes just the act of typing words is all it takes to get started.

  Rather than get stuck again, just keep typing. Don't try to stop and revise and obsess over your wording—save that for later, when you review. As you continue, you'll eventually start to type fresh content in your own words.

- **Dare to be bad**. Maybe you're afraid to write until you are confident it all will be completely accurate. Forget that! Write about something the way you think it works, even if it might be wrong. Be liberal with your author's comments and questions so you know where to question a subject matter expert later, and make sure those comments stand out so you don't accidentally let incorrect material go into the final version.

  When you read your "bad" paragraphs later on, you might discover to your surprise that they aren't so bad after all. In fact, they're probably pretty good. Things look a bit different when you give them some breathing room. But even if they're not the best, they have some use. They can be reviewed and revised and they will make up a part of your final content. Edit what's usable, delete what's not, and pat yourself on the back for daring to be bad. Move on, because now you're rolling.

- **Make an outline.** (Not *again!*) I know I told you that already. But if you've got that bad a case of writer's block, I'll bet it's because you didn't make an outline. Or if you did, it's so skimpy it's not of much use. Go back to review the steps on making an outline, and make it happen.

I'm betting you won't be terminally blocked if you use the tactics in this chapter. After all, you're a writer, and writing is what a writer does.

Chapter 16

Everybody's a Critic—Reviews and Reviewers

The importance of getting your work reviewed, and how to go about it.

What's in this chapter

- Finding—and keeping—reviewers
- Different ways to hold your reviews
- How to incorporate feedback
- Saying "thank you"

Worse than everybody being a critic is when nobody is—when you work hard on your documentation, send it out in email, and then hear...nothing.

Stakeholders—the people at your company with a vested interest in your project— are busy people and they don't always have time to read documentation. It can often be difficult to get their important feedback and to get it in time to meet your schedule.

In this chapter, I'll guide you through the ups and downs of this important part of a tech writer's work life.

Reality Check

As I've said before, I believe that a technical writer should be able to write documentation and feel confident about it going to customers even if no one has reviewed the material. If a tech writer understands the product well and understands the company's objectives, there is no reason why this can't be the case.

However, we're not all perfect and we don't know everything, and it is good to have other pairs of eyes on our work. Reviews can be eye-opening. You will discover that you left out key information, that the procedure you thought was perfectly accurate actually has mistakes in it, and that your spell-checker let "kick" go through instead of "click."

So it's still advisable to strive for a situation in which you understand things so well that you are capable of writing without reviews (and this will also save your reviewers time if your drafts are that good). And then, to make it even better, you should have reviews and sign-offs anyway.

Keeping Current

It's important to get confirmation that what you've written is indeed accurate and complete—and current. "Current" is the issue that can elude you. It may be the case that there is old, out-of-date information in your documentation and you haven't noticed it.

Or, it could be that changes were made at the last minute or without your knowledge. It might be because you missed the hallway conversation where one developer told another about changes made to the software packaging. Whatever the reason, what was true and complete when you wrote it is suddenly not so true and not so complete.

The review process is the only surefire way to catch these issues. (Well, almost surefire—those same people who neglected to tell you about the hallway conversation might also neglect to read your documentation.)

Filling in the Gaps

Figuring out what's missing is not always easy.

A reviewer can look at something, see that it's essentially correct, and not even think about information that's *not* there. But if the reviewers are on their toes, you'll hope a sharp-eyed one will see what's missing and tell you where it fits.

Getting Sign-off

You don't want to be left holding the bag if a problem or oversight happens after your documentation goes out. Things happen whenever anything is released to a customer, and fingers can be pointed in a very unpleasant way if you let documentation go out without reviewers seeing it. It's a good idea to make sure that the product manager is on board with your documentation, and the engineers who developed it agree that what you've written is correct.

Designate someone early who will be responsible for signing off on the documentation. (This is usually the product manager.) A sign-off doesn't have to be fancy or require a notarized signature. All you need is an email message in which the product owner says the deliverable is approved for release.

It's Not Personal

With all these reasons to have documentation reviews, you can see how important it is to make sure everyone has the same goals, and that includes you as well as your reviewers. Let go of any emotional investment you might have. This is not the product of your blood, sweat, and tears (I hope!)—it's technical documentation. It's your work output.

Yes, it should be as good as you can make it. You accept a paycheck from your employer and it's your responsibility to do the best you can. But a critique of the document is not a critique of you as a person or even as a writer.

A surprising number of tech writers haven't grown this fundamental "thick skin." You simply can't take it personally when reviewers start cutting apart a document you've worked hard on. And cut it apart they will; it's what you asked them to do. You want them to push it, pull on it, point out its weaknesses. That's the only way you'll know how to make it better. Even if their remarks also become cutting, which sometimes happens, you've got to shrug them off.

Some reviewers have had experience with overly sensitive technical writers and are afraid to say anything that can be seen as criticism. Make sure you're not one of those writers. Invite constructive feedback; it's the only way to improve your work.

Choosing Reviewers

Choose reviewers whose input will be meaningful. Usually this means the people who helped provide the information that's in your documentation. This means everyone you interviewed, who gave you demos, and who answered your questions as you wrote the first draft.

Think also of whether there are people in other departments who need to be aware of the documentation, or who might have unique insights to offer. This can include people in Sales, Marketing, Customer Support, or Operations. If you aren't sure if you covered everybody, put "Please forward this to anyone I may have missed" in the email that contains the review announcement.

Make it very clear exactly what you need reviewers' help with: content, not grammar and syntax. Some reviewers annoy the writer by spending time and attention on punctuation, style, and grammar.

Make sure you are scrupulous with the spell-checker, because it can take only one typo to convince a would-be editor that you really do need help with writing.

For a first draft, you don't usually need to send it out to management, unless someone like your manager has explicitly asked you to. At the first-draft stage, you are still trying to make sure the information is correct. (Of course, if this is the only draft you intend to send out, make sure *everyone* sees it.)

Be sure to let your reviewers know ahead of time that the review is imminent. A review can take a significant investment of time, and people don't appreciate being surprised. Make sure that your reviewers are able to fit this within their schedules—if they are in a time crunch, working on a different deadline than you are, they are going to have a difficult time doing what they see as *your* job.

This is where Agile methodology, discussed in Chapter 9, "Process and Planning," can be very beneficial. In an Agile environment, writers and developers should be in sync, working in sprints to develop both the features and the documentation on the same schedule. Reviewers look at only a couple of segments of documentation at a time while those particular topics are still on their minds, and at the end, you can assemble it with the knowledge that it is correct.

Anyway, that's the way it's supposed to work. Nothing is ever really that easy—changes to software are made on the fly even though development was supposed to be frozen, transitional information must be added to glue all the pieces together, and the grunt work of preparing manuals and help and web content for production has yet to happen—but I still think that Agile and the scrum system is one of the better ways for getting your content reviewed.

Educate the Reviewers

Sometimes, it's a good idea to educate reviewers before that first review. There are three things reviewers might not know, but need to. It's probably up to you to tell them.

- **About the documentation plan**. Many developers have no understanding of what happens before, during, or after the review process they participate in. It helps them to help you do your job if they see how their part fits into the big picture. Make sure everyone on your review list has seen the documentation plan.
- **They can ignore typos, grammar, stylistic issues, and personal preferences**. Hopefully, you addressed any potential problems by putting out a good solid draft that's been given a spell-check and a once-over. Ask your reviewers to focus on the correctness of the content and not be distracted by grammar and syntax or whether they like the formal passive voice better than the second person.
- **It's OK for them to tell you what's wrong with the documentation**. Get your hard shell on and tell your reviewers not to hold back for fear of hurting your feelings. If you can take whatever they can dish out, it will be better for your work in the long run.

Sending the Draft Out for Review

Different writers handle reviews differently. Some writers believe in many short mini-reviews and question-and-answer sessions with their co-workers, reducing the need for repeated reviews of the whole content.

Other writers believe in a larger collaborative team effort, and depend on subject matter experts to read the entire documentation deliverable and provide extensive and detailed review comments throughout multiple review cycles.

Whatever your preference, the feedback process is important. It provides expert reviews to ensure that the material you write and assemble is correct. It also lets product owners and other stakeholders see and sign off on the messaging that goes out to customers.

Making the review process easy for reviewers goes a long way toward getting the feedback you want and need. As discussed in Chapter 15, "Putting It All Together,"you want to make sure that this content is clearly marked as a draft and is as final and neat as it can be. Make it look as if you care, so your reviewers will respect your work. Don't distract them with typos, mistitled cross-references, incorrect information, and gaping holes of missing content.

Turn On Those Change Bars

The moment your documentation goes out for review, turn on the change bars so that when the reviewers see your *next* draft, they will know exactly what has changed.

Change bars let your reviewers know what's different between the previous version and the one they are reading now. When reviewers see the second draft, they can save time by looking only at the areas that have been changed.

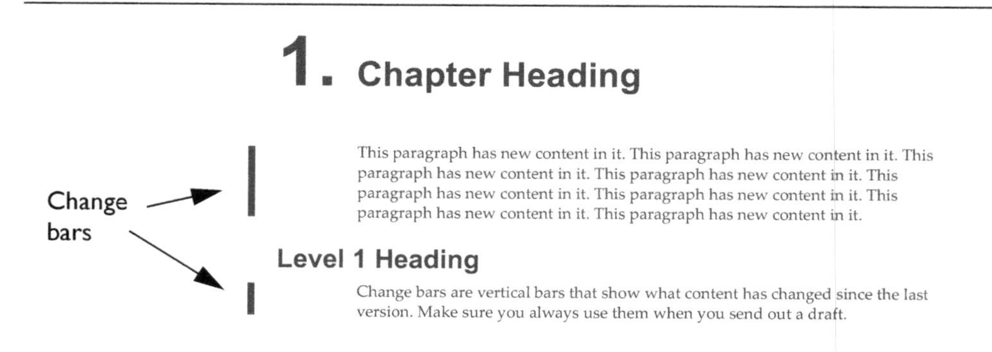

Change bars in the margin highlight content that has changed since the last round.

Make It Small

When your colleagues receive very large amounts of documentation for review, they tend to set it aside for when they have time. If the documentation is too big, sometimes that time never comes.

To get a better response to your review requests, treat your reviewers like your most important customers by customizing your documentation for them. Send each reviewer the smallest chunk of documentation you can and don't make them look at sections that don't apply. Being faced with 100 help topics can be pretty terrifying for a reviewer to think about, but if it turns out that only 20 of them relate to his area of expertise, he probably doesn't want to read them. Be nice and carve them out so he sees only what he needs to see.

It often happens that a writer working on multiple deliverables performs the final touches on them all at the same time, and within a couple of days, sends out not one, not two, but three or even four different documents for review.

Think about how that feels to the reviewer and make sure that only the people who need to be on the review list are. Work out a staggered schedule with the reviewers whose input is needed on more than one item. If one of your deliverables is a help set that shares content with a user guide, don't send out the help for review, or at least wait until after you get feedback on the user guide.

Set Expectations

When you send your draft out for review, be very clear about what you expect from your reviewers. If there are any points you need them to look closely at (or disregard), tell them now. Make sure your author questions and comments are clearly defined in the review copy so no one can miss them.

Set a Deadline

Too often, writers forget to include a due date with their reviews. Make sure you are clear about when you need the review comments back.

Deadlines are tricky. If you give your reviewers too little time, they will be (rightfully) angry at your lack of consideration. If you give them too much time, they tend to put the job off and forget about it. Three to five days is usually a reasonable amount of time. Under no circumstances should you ask people to turn in comments within 24 hours unless it has already been declared an emergency by the product team.

Look at the calendar before setting the deadline and make sure you take weekends and holidays into consideration. Monday is not a good day to set as your deadline, because even though many people do work over the weekends, most will not, and it's another opportunity to forget your review. The end of Thursday is an ideal time for review comments to come back, because it gives you one workday to either remind the recalcitrant reviewer or to incorporate the comments. If you can, ask for comments back a bit sooner than you really need them, so that if a key reviewer does not respond, you have some extra time to chase down that important feedback.

Let's face it—no one likes to be given two days to read a document the size of an old-fashioned telephone book. But experienced tech writers know that if they allow too *much* time, the reviewers will wait until the last minute or forget that they had this task to do.

Send a Reminder

Feel free to remind your reviewers as the deadline comes closer, either in email or in person. If your review copy is sitting in someone's mailbox, it's far better to remind her of the deadline today than to feel annoyed tomorrow when she doesn't return it in time.

Continue Working

While your documentation is out for review, you should continue working on it—to fill in the content you said you would supply (remember all those TBSs?), to make sure you get some of your unanswered questions answered, and to work on final touches such as formatting, indexing, and hyperlinks.

Gathering Review Feedback

There are only so many ways to collect review comments from people. You want a system that lets reviewers easily insert their comments and corrections in a way that is visible to all reviewers. (This prevents people from wasting their time repeating comments that were already made.) You would also like to make sure that reviewers address *all* of your questions and gaps.

At the time of this writing, there doesn't seem to be a perfect tool for reviews, but there are several methods—some manual—that may work for you. Depending on what is most important to you during this phase of your review, you can choose the method that is best.

Email Distribution

The simplest and probably most common way to distribute documentation for review is to attach it to email, in a PDF or Word file, and ask the people on your distribution list to return comments by a certain deadline.

Reviewers will provide feedback in a number of ways:

- Typing out their comments and returning them in email
- Revising the file itself by adding or inserting their comments and deleting text that is not relevant
- Telling you verbally what they think needs to be changed
- Marking up a paper copy with a pen or pencil

Collaborative Reviews

Letting reviewers collaborate and see each others' comments is, in my opinion, the best way to handle soft-copy reviews. Reviewers type their changes into their copy, and the changes are merged into a single file that everyone can see. The advantage to this is that one reviewer sees what another one says and can disagree if necessary. In any event, it saves the time others would have used to type the same thing. This also puts all of the feedback into a single file for you, the writer. Instead of working with multiple copies in every kind of format from FrameMaker to Word to paper to email, you have all feedback consolidated.

Web-based spaces like Google Docs or Microsoft SharePoint permit many people to write to the same file. A user checks out a document to save changes to it, and checks it in when done. The user's changes are visible to all who have access. If there are any problems with a particular check-in, you can roll back to a previous version.

Wikis enable collaboration, and developers seem to like wikis. Ask your software development group if a collaborative environment exists right now in which you can post review documents. If you don't have anything like that available at your company, look into setting up a simple one yourself.

Adobe Acrobat's Shared Review feature lets you share PDF files with reviewers. It turns on editing tools so reviewers can mark up the PDF and everyone can see the changes. You can send the review copy in email, the reviewer works in the attached file, and—voilà—each reviewer's published changes are seen by all other reviewers when they open their own copies. Changes are stored for each reviewer on a drive within your company or even on the Adobe site. Because your company's drafts should be kept confidential, it's best to use a corporate drive. Discuss this with IT and make sure the drive is open to everyone within the company who might ever be a reviewer (this might mean everyone in the company), so their changes can be saved to it.

> **INSIDERS KNOW**
>
> At a tabletop review, see if you can get one of your Tech Pubs colleagues to act as the moderator while you take notes and ask questions. This method allows you to concentrate on collecting the important information with no distractions. To make sure you collect all written feedback as well, ask reviewers who have marked up paper copies or taken notes to let you photocopy or keep their notes.

Tabletop Review

If you are developing a great deal of new content, a tabletop review is one of the most useful ways to get feedback. Tabletops are also good for documents that generate a lot of interest due to their importance to the company or because of some sensitivity about the messaging.

In a tabletop review, all the reviewers gather in a conference room around a table (hence, the "tabletop") and together go through the document page by page, paragraph by paragraph, even word by word if necessary. You can project the material on a screen, if it is readable, or even pass around hard copies if it is very text-heavy. Everyone's feedback is shared with the group and any disagreements or clarifications can be handled as they come up. You, as the writer and review facilitator, keep the focus on the document and take note of what changes need to be made.

Facilitating a tabletop review does take focus. And a firm hand. You need to make sure that the discussions do not go on forever, or you never get past the first page! Most, if not all, of your reviewers will not have read the material before the session, and will be scanning it during the meeting. They are bound to be surprised by what they see.

You might need several hours or more for a full tabletop review, depending on the length and complexity of your documentation and the number of reviewers involved. However, it's unlikely that people will commit to more than an hour or two for a session.

One solution is to break your review into smaller sessions of one to two hours each. For example, you could plan to go over one subject in the first hour, another subject in the second hour, and so on. Schedule the room for the entire day, or however long you think it will take. Reviewers can drop in for their session, and if they are interested, they might stay longer.

Most reviewers would like—and ask for—a copy that they can mark up. If you work in Microsoft Word or a tool such as Adobe RoboHelp or MadCap Flare that works well with Word, this is easy. If you work in FrameMaker, be willing to convert your documentation into a format that reviewers can easily work with. A reviewer can turn on Track Changes and happily edit, and you can incorporate the changes as needed.

Especially if you provide lunch!

One-on-One Review

For reviewers who are really important sources of information, I suggest you schedule one-on-one reviews with them independently of the group reviews.

Ask the reviewer to read the documentation first, if possible, so when you meet, you will be able to maximize your time together. If the reviewer was not able to read the document, don't let this stop you. Have important passages highlighted and ready to review, and have the reviewer read them on the spot. Have a list of specific questions ready, and make sure you get them answered.

Ongoing Mini Reviews

I believe the most effective way to get feedback is to have regular and ideally in-person discussions with the people who hold the information you seek. Instead of churning out 2,000 words, go ahead and ask someone to look at the 80 words and diagram you just created and let you know if it's correct. Getting that immediate feedback can save a lot of work for everyone later—no one really wants to read a massive tome—and if you know your material is accurate as you write it, it keeps you on track for making the finished product accurate.

Consolidating Reviewer Feedback

While you continue to work on your documentation, some (all, if you're lucky) of your reviewers are reviewing your work. When the deadline has passed, you will have comments in a number of places—the shared review copy, in email or any of the other ways we talked about. It's your job to collect all the comments and incorporate them into your document.

It's best to review all feedback before you start to incorporate it. The reason for this is you don't want to get into a situation where you make a change suggested by one reviewer, then discover that another reviewer has a completely different idea. You incorporate his changes instead, and suddenly a third person speaks up in email with a different set of suggestions in the same area, and you realize you've lost control of your project.

Once you've read and digested all the reviewer feedback, you should have a pretty solid idea of what you will change. The best way to incorporate reviewer comments is to systematically go through each comment one at a time, and either type the change into your master file, reject it, or get more information. If you are working from an edited PDF or Word file, you might want to print out the file so you can check off each comment as you address it. (If you *don't* check off comments that you deal with, you are sure to miss some important ones.)

Sometimes it happens that two reviewers make statements that flatly contradict each other. Work with both of them to resolve this situation and if necessary, let them duke it out and let you know what conclusion they reach.

Once More, with Feeling

After you've collected all the feedback, resolved reviewer conflicts, and incorporated changes and additions into your documentation, you're ready to...do it all over again. Yes, again. With a slight difference.

In your second draft, you want reviewers to focus on what has changed since the first draft. It is helpful to state in email what the broad changes were, so the reviewers know what to expect. Make it easy for them by using the change bars mentioned previously. A reviewer should never have to read everything again to try to guess what is different or whether her changes were included.

Reviewers don't like to read the same document over and over. If they see it a second time and it still has major problems—or worse, the *same* problems—they are resentful and feel as if they are doing your job. Even when the draft doesn't have repeated problems, it's hard to get as many enthusiastic responses the second or third time around as you did the first.

Make sure that when you send out a review, you have addressed changes from all of your reviewers. And by addressed, I don't mean you have to *make* the changes, but you at least must look at them and make a decision as to whether they should be implemented.

You can't send out a new draft to a reviewer who gave feedback if you have ignored her feedback. It's not a valid excuse to say you ignored a reviewer's comments because the reviewer submitted changes after the deadline. If you do this, you are asking the reviewer reread everything only to discover that her input was not included.

If you have any pending feedback that has not yet been dealt with, make sure you address this fact before sending a draft to all. If you can safely delay your review, go ahead and do that. If not, contact the reviewer whose feedback is pending and explain what has happened. Give her the option of looking at the changes that have been made already by others, or waiting until the next draft.

How many repetitions of this review cycle do you need? The theoretical answer to that question is "however many it takes to get the content right." But the practical answer is "not more than one or two." If you've done your homework, gathered all the information, organized it lucidly, talked with the right people, and solicited and incorporated feedback from knowledgeable people as the questions come up instead of waiting until draft time, you shouldn't need more than two review cycles. Which is a good thing, because it is unlikely that you'll have time for more than two, anyway.

Change Management

Of course, life in high tech doesn't always go like clockwork. The product can change radically, the people involved in a project can change, an internal process can change, even a company's business objectives or marketing positioning can change between the time you send out your first and second drafts. They can even decide they need a totally different piece of documentation, or even cancel the product, at the last minute. These changes occur all the time and it's your job to roll with the changes.

I know it can be frustrating to have to incorporate these changes, especially when you discover that major decisions were made and no one bothered to tell you about them. Take it in stride and think of this as part of a tech writer's job, not something that's been done to make you unhappy. If you can learn to laugh, you'll be able to survive anything!

Thank Reviewers for their Contributions

Don't forget to thank the people who took the time to give you comments on your draft. Even if it's part of their job (or you think it's part of their job), it's a lot of work to read and correct someone else's work. Be grateful for their contributions, and do what it takes to assure that the next time you have documentation to review, they will cheerfully participate.

Chapter 17

Wrapping it Up

Finalizing your technical writing project.

What's in this chapter

- How to polish your documentation till it glows
- Tips for editing and proofreading
- Wrapping up with a final document checklist
- How to make a document freeze not feel like a cold shoulder
- Making sure the customer sees your documentation

At some point in the review cycle, you'll feel confident that the draft you send out for review is indeed the last draft. You're making the final laps leading into the home stretch. Just a few more times around the track for final editing, proofreading, and testing, and you'll be able to get final approvals and sign-offs and wrap that documentation up for customer delivery.

But the last few laps of a race can always hold a few surprises. This chapter helps you to be prepared and ward off those surprises so you can finish up with no problems.

And when you finally hand off that deliverable, you can be assured that it's the best it can be and you can congratulate yourself on a job well done.

Testing, Testing...

Until someone actually reads and uses your documentation, you don't really know how good it is. If you are lucky, your company's Usability or Quality

Assurance (QA) department provides some document testing. During this phase, the tester should go through all the procedures and confirm that they work, and check to see that screen shots match the software. When testing online help, the tester checks to see that all the links work as expected.

Unfortunately, not every company incorporates a document-testing phase into the release cycle. Tight deadlines and overloaded schedules in the high-tech world make it a rare luxury.

If you don't have the benefit of QA testing (or even if you do!), make testing a priority by doing what you can yourself or asking your colleagues to help. Follow the procedures in your documentation, step by step, and make sure you haven't made any errors or left out any crucial steps .

Testing is particularly important for installation documentation. Ask someone who's unfamiliar with the process to go through your documentation and perform the installation procedures. To test online help, make sure all of your hyperlinks work and lead to the intended place.

Rewriting and Editing

If your working environment is like that of many technical writers, you'll be expected to manage your own editing and proofreading. Full-time, on-staff editors are a luxury many companies don't see the need for (if there is an open requisition for a new employee, Technical Publications hiring managers tend to fill it with a writer rather than an editor). If you are lucky enough to work for a department that has editors or proofreaders, be grateful. If not, read on.

Insiders know that the final stages—editing, formatting, running a table of contents and generating into PDF—take much longer than expected. Allow plenty of time for cleanup and be prepared for surprises. You wouldn't believe how often a snag occurs right at the end.

As you went through the review cycle, you probably spent a lot of time adding comments and revising content. Every time you touch your content to make it better, you also run the risk of introducing a careless mistake. Your content should be solid by now, but it's a good idea to take a final pass from a strictly editorial standpoint. There should only be minor editing at this point, but sometimes you'll find more than you bargained for.

Hard as it might be to believe, it can take more focus and discipline to rewrite and edit than it does to write raw content. When you were composing your drafts, there was a certain freedom—it was a discovery process as much as a construction process, and you rearranged as you went. You also might not have paid sufficient attention to consistency and correctness in your references and terminology, or followed the best practices as closely as you should.

Proofreading and editing require an eye for detail. You must bring a critical eye to your writing, and make sure the rough spots you let slide earlier are polished to a high sheen now. You must look at the project as a whole and be rigorous about the parts that make up the whole.

How Much Is Enough?

It's a judgment call about how much rewriting or editing to do. There are *always* improvements to be made. But if you're on a tight schedule, you probably can't do everything you'd like to do, so you have to pick and choose.

Let's go back to the fundamentals described in Chapter 5, "How to Write Good (Documentation)." There we learned that documentation should be correct, complete, usable, clear, and consistent. Think about all of these criteria as you perform your final editing and proofreading tasks.

Although the review process should have ensured that the content is correct and complete, that doesn't mean that all the details were covered during that phase. I can't tell you how many times a document has been handed off and published or publicly posted with incorrect headers on the pages, typos, formatting errors, author's comments and draft markers left in, and references to other documents with incorrect titles. Those are all mistakes that should be found and corrected during the editing phase. Make these corrections your top priority.

Cast a keen eye on anything that affects the clarity of the documentation. If you don't understand what a sentence or word means, that is a top priority. If you see any passive voice that adds to ambiguousness, change it to active. Is there a

In 2008, a group of copy editors who call themselves the Typo Eradication Advancement League (TEAL), went on a cross-country trip in 2008 to correct typos across the United States. They corrected two typos on a sign at the Grand Canyon and discovered later that they had vandalized a historic marker hand-painted by Mary Colter in 1932.

They had to pay $3,000 in restitution and were banned from national parks for a year, resulting in a new TEAL policy of always asking for permission before they fix a typo.

Read more about it in *The Great Typo Hunt: Two Friends Changing the World, One Correction at a Time* by TEAL founders Jeff Deck and Benjamin D. Herson.

single, clear idea in every paragraph? Split up any overly large paragraphs into two. Is there a clear lead-in for each bullet list or procedure? Add one if not.

You might look at a sentence or paragraph and have a vague sense that it could be better, but you can't put your finger on exactly what needs to change. Leave it alone. If the problem isn't immediately obvious to you, it probably won't stand out for the reader either. Spend your time on issues that need more attention.

Remember the importance of consistency and give it the attention it needs. Make sure you use the same terms to mean the same thing every time, that the structure is the same throughout the document, that lists have the same structure and punctuation. Refer to Chapter 5, "How to Write Good (Documentation)," if you need a refresher.

Keep Things Moving

Keep it simple and remember that a light touch is better when you edit. Once you start doing major rework or revision, it's easy to become bogged down. If you think you might really need to spend a lot of time on a certain area, make a note to yourself to come back later. Move on and return to the problem area after you've been through the whole project. Your drive is for a consistent level of quality, not a mix of diamonds and coal.

If you have time after your first pass, go back to the difficult parts and address them in order—first, according to their importance to the document and second, according to how quickly and easily you can correct them.

Editing Your Own Work

It's much harder to edit your own work than someone else's. Why? Because when it's your own work, you know what you meant to express, and you might not realize it if you fail to accomplish it. And you become so accustomed to gaps and flaws in the content, you don't see them anymore.

Simply put, it's difficult to be objective about your own work. Objectivity comes a lot more easily when material is brand new to the reader. You just might be tired of looking at your document by this time!

Be aware that when you edit your own work, you will not do as good a job as someone else will. That doesn't mean you shouldn't do a final edit. I think it's important for writers to give a final read to their own work and do the best they can to clean it up before it goes out.

To do the best editing job on your own work, set it aside so you can gain some distance from it before you look at it. The longer the better, but even a day's distance helps.

It also helps to break the editing process into several sessions if you can, because once you start reading your work, you become close to it again. You might start to gloss over it without making any changes. There's a difference between the deciding to make minimal changes during "editing triage" and simply not seeing what needs your attention. Ask yourself if it's really that perfect, or you've just become numb. If the latter, it is probably time for a break to regain some distance.

Editing Another Writer's Work

Tech writers often perform peer editing for each other. This brings a fresh pair of eyes to the content, one that understands the company and also the templates and formats that should be used. Your tech writing peer might also be familiar with your product, which helps a lot. If he's not, he'll learn something, so if he ends up working on the project later, he will have some familiarity with it.

Proofreaders often concentrate so closely on the body content, they miss big errors in the titles and headings. There are many horror stories of newspapers, books, or posters going to print with a big embarrassing typo in the main heading. Don't forget the headings when you review your work!

Although it's much easier to do a good job editing what someone else wrote, remember that this is not a free-for-all. You cannot rewrite with the same abandon you might apply to your own work. Keeping it simple is more important than ever now. In short, you must remember to have consideration for your colleagues when you edit their work. Treat them and their work with respect.

You must accept that your colleague has a different writing style than yours, and not try to enforce your own style when editing another's work. If the content is clear and no department style guidelines have been violated, then the writing is legitimate and does not need to be revised. If the content is not clear, ask the author what was intended.

As an editor, keep your attention on the five important attributes of good documentation: correctness, completeness, usability, clarity, and consistency. You can let go of anything else.

Proofreading

Are you one of those people who drives your friends crazy by pointing out typos in everything from store signs to restaurant menus? Does it bother you at

If proofreading is something you struggle with, there are strategies you can use to get yourself through it and do a better job. One of them is, believe it or not, to read the document backward, word by word or paragraph by paragraph. Start at the last page and work your way back to the front. This puts words out of sequence so you don't associate meanings with them.

Another tactic is to put a white card or piece of paper across a hard-copy page underneath the line you are reading. Move the paper down the page as you read. This slows you down and causes you to focus on each line, helping mistakes jump out more easily.

If you are reading on screen, zoom in closely on the text so you read only small portions at a time. Like using the piece of paper below a line, this isolates sections of text to allow you to focus on the details and not on the meaning of the content.

a visceral level when you see an apostrophe before a plural "s" or the word "your" when the writer meant "you're"? If that sounds a bit like you, you probably won't have any problems proofreading.

Proofreading means reading through a document and paying meticulous attention to fine details: spelling, punctuation, grammar, capitalization, correct use of typefaces, the right titles and captions for illustrations, line breaks and page breaks, even the spacing between headings and paragraphs. Think of proofreading as going through content with a very fine-toothed comb.

By the time you are at the proofreading stage, you should not be thinking about content or clarity or correctness. You're no longer wondering whether a figure clearly illustrates the concept, but only if it is aligned on the page according to the style guide or if the figure caption is spelled right and capitalized according to your department's standard.

I recommend that as part of your proofreading task, you convert your documentation to its final format (PDF, HTML, or whatever you will deliver) and print out a hard copy of it. The PDF conversion sometimes includes font changes that can make minute differences between your original source file and the final output. And I guarantee you will see mistakes on a paper printout that you never notice on screen.

You don't have to mark up your copy the old-fashioned way, by using a red pen and proofreader's marks. It's probably best, and fastest, for you to skip the markup stage, visually refer to your printout, and when you find mistakes, make the corrections directly into the file.

But proofreading with a pen or pencil does help improve your accuracy. Marking a hard copy and then typing the changes into your electronic files helps avoid the problems that occur when you start freely changing text and suddenly realize it's gone in a direction you hadn't intended.

And marking up a paper copy is a very portable activity—sitting with a stack of paper on your lap is much more civilized than typing away on your laptop while at a meeting or a presentation. It's an easy job to take with you on the train ride home or while eating lunch or watching television.

Here is a table of proofreader's marks that have been used in the editing field for years.

Mark	Meaning	Mark	Meaning
	Delete	sp	Spell out
	Close up (delete space)	A	Use lowercase letter instead of capital
#	Insert space	a	Change to capital letter
	Start new paragraph	bf	Make bold
	Move to left	a	Make italic
	Move to right		Insert comma
	Center		Insert apostrophe
	Transpose		Insert period
stet	Ignore my markup; let it stand.	wf	Wrong font

A Final Delivery Checklist

As a last step before releasing your documentation to customers, it is a good idea go through a final checklist and make sure everything is done. The final checklist ensures that you catch those little embarrassing things that are too often missed. If your department does not have a standard checklist, create one yourself and share it with your colleagues. Here are some items you should include; use those that are relevant to your project:

- Run the spell-checker.
- Apply the latest templates.

- Clear all change bars.
- Remove draft markers.
- Remove all author's comments and questions.
- Make sure chapter numbers and page numbers are sequential and correct.
- Make sure all links work.
- Update table of contents, index, and cross-references.
- Check that pages break correctly.
- Delete empty pages.

Freezing the Documentation

You might expect that you would have stopped making changes to content long before you reach the proofreading stage. It would be nice if that were true, but the fact is, things don't always work that way.

At some point, you must shut the doors on any more proofreading, revision, or incoming changes if you are to meet your deadline. This is called "freezing" the documentation. Make sure the product team is aware of your freeze date (or freeze time, if you are cutting the schedule that closely).

When your documentation is incorporated into a software build, put on some kind of media (such as a CD or DVD) that goes to the customer, or printed, it's important to stick to your freeze dates. If your documentation is relatively easy to update—on an extranet or a corporate website—you can be a bit looser with your freeze dates, based on the fact that you can make post-release revisions.

No matter how perfect your final documentation looks in Word or FrameMaker or RoboHelp, always look at it in its final form. If you are producing a PDF file, print the PDF and page through it one page at a time. I guarantee you will find odd page breaks and typos you never knew were there.

However, even if you can make post-release revisions, that doesn't mean it's a good idea. An eager customer will download the documentation as soon as it becomes available. Subsequent changes often don't even reach that customer.

And if you feel free to make regular changes to documentation, you might not take the first version seriously enough. The version you first release to a customer should be as complete and correct as it can be, to the best of your knowledge. Plan to make revisions only in the case of an emergency.

Will requests for changes continue to come in after the freeze date? Of course. But usually, changes can wait until the next document revision.

Collect and organize change requests so that when you are ready to update, you know exactly where all proposed changes are. Keep up with engineering schedules so you can plan to do a revision when the software or firmware is revised—at least then, you can assume the customer might look at new documentation. Assess the requested changes, prioritize them, and be firm, but friendly, as you agree to do changes at a later date.

Hold that Freeze!

There will inevitably be a time (many times) when any notion of a solid freeze date goes out the window. Like it or not, freeze dates are more often slushy than they are rock-hard.

Just as you're ready to hand off the final documentation, you're likely to see something—as if for the first time—and realize it is out-of-date or just plain wrong. A product manager might email you and tell you something must be added to meet contractual obligations for a customer. Or you might happen to learn that a feature has not made it successfully through QA and is being removed until it can be corrected for the next release. These and hair-pulling events like them happen all the time in the world of technical writing.

In general, all you can do is accept the changes—with a smile, if you can muster one—and scramble to get them in. If your documentation changes affect a software build schedule, discuss the revised schedule with the broader team, and make some decisions about what to do. You might end up working all night, or the team might decide that the changes aren't that critical after all. Or you might realize they can go into a release note.

Sign Off

Find out if your department uses a formal sign-off process, and if so, follow it. If not, cover your bases by emailing a clean version of your documentation in its final form (or in whatever form your reviewers can read) to all of the project stakeholders. Make sure you include your manager, the product manager, and any other management types who are interested in the documentation. It's a good idea to include Customer Support, too, so they have the latest version of the product documentation.

Make sure you tell everyone this is the final version. If there is still time to make emergency changes, let them know, but it is probably even more important to tell them when it's *not* possible to make changes. In that case, tell them when the next opportunity for change will be.

Send Off

The last step to finalize your documentation is to hand it off. As we discussed in Chapter 12, "You Want it *How?*"its final residence is not on your hard drive or in your version control system—it's got to go live some place where a customer can access it. Find out what its final destination is.

Typically, this can be any of a number of places:

- Company extranet (a private network that allows controlled access to partners or customers)
- Company Internet
- Internal wiki or Microsoft SharePoint site
- The software build, so it is accessible from the application's user interface
- On an electronic device

After you deliver the documentation, as a final step, put the finished documentation in your version control system or directory or wherever documentation files reside, and create a new set of files for the next version. Now you can finally say you are finished.

Can you believe it? You've completed the entire product lifecycle for documentation. You took it from concept to draft to final production.

Put your feet up for a few minutes and give yourself a big pat on the back. You earned it.

Part 5. The Tech Writer Toolkit

This section is like having a staff of consultants on hand to help you with the entire infrastructure it takes to build a Technical Publications department. Here you'll find not only descriptions of style guides, indexes, book design, clear writing, and localization, but also recommendations and examples you can use to create all of these aids yourself.

Chapter 18

The Always-in-Style Guide

Consistency is key to finalizing your technical writing project.

What's in this chapter

- How a style guide can be a time-saver
- The problems style guides solve
- Recommendations for your library
- Creating your own style guide

We've spent a lot of time learning about how to gain expertise in your subject matter, how to gather information from others, and how to use the tools that help you create documentation. Those tools are indeed important for forging your content. But crafting your documentation takes something more—it takes style.

I'm not talking about how you wear your jeans or whether there's a designer name on your laptop bag. For a technical writer, *style* means doing things in a particular, deliberate, and consistent way according to rules that govern everything from how to capitalize titles to how to write numbers. The collections of these rules of style are called, not surprisingly, style guides.

Without a style guide, writers can, and will, choose any way they like to treat capitalization, punctuation, and even spelling. Taken to an extreme, this type of inconsistency can hurt the company's brand—it looks unprofessional when names and terms are inconsistent across a company's products, especially when product names are wrong. It can be a big time-saver when you don't have to think or wonder about how to treat bullet lists or numbers or other style issues.

Is there more than one right way to do things? Of course. And it doesn't matter which right way you select, as long as you, and your fellow technical writers, pick one and follow it consistently. That can sometimes be hard, because we all have our favorite ways of doing things.

What Makes a Style Guide Such a Hot Property?

I've recommended that when you start work at a new job, you should ask for the company or department style guide before you even ask for your first cup of coffee. Yes, a style guide is really that important.

That's because a style guide, if done well, contains the answers to the infrastructure of the documentation and the answers to nearly every question a writer might ask about how writing should be done. Once you understand the style guide, you can concentrate on content.

The term *style* refers to two different aspects of writing: mechanics such as word choice, phrase use, punctuation, spelling, and other usage conventions, as well as more abstract issues such as tone and structure. A style guide should address both of these.

By setting clear standards, a style guide enforces consistency within an individual deliverable as well as across the company's products. Consistency, as we've discussed, is essential to effective technical communication.

A style guide presents guidelines for many topics, usually including:

- Abbreviations and acronyms
- Capitalization
- Consistent word usage
- Product names and trademark usage
- Punctuation
- Spelling
- Tone
- Word usage

Style guides may also contain typographic, formatting, and layout guidelines. If you decide to include those in your style guide, refer to Chapter 20, "Design and Layout," for details.

Style Equals Speed

At first it might seem as if using a style guide will slow you down. ("Oh, no, I have to look something up again instead of doing it the way I prefer!") but the truth is that once you learn the accepted styles, you can actually work faster.

One of the interesting—and frustrating—features of the English language is that it offers many ways to express essentially the same idea. For the fiction writer, this provides great richness and texture, but for the technical writer, the wealth of choices can slow you down.

The clock is always ticking for the technical writer. Do you really want to spend even five seconds of your time deciding whether you should hyphenate "end user," or whether "backup" should be one word, two words, or hyphenated? How much more efficient you would be if you could focus only on content instead of worrying about these other details.

Where a style guide helps you pick up speed is by saving those precious seconds you spend every time you pause to think about usage or phraseology. Without a style guide, you might easily do something differently every time.

> **COFFEE BREAK**
>
> One space or two? Many writers argue about whether to put one space or two after the end of a sentence. Typing two spaces after a period is generally agreed to be a holdover from the days of typewriters. With typewriter fonts, two spaces made it easier to see where the sentence ended and the new one began. Typesetters, however, using proportional fonts, put only one space after a sentence.
>
> With today's proportionally spaced fonts available to all, we are all typesetters and there is no need to use two spaces. People who have learned the two-space rule have a hard time breaking the habit and often argue that it is better for readability. *The Chicago Manual of Style* believes there is no reason for two spaces after a period in published work. I hope this is not an ongoing argument in your Tech Pubs department! There are so many better things to do with your time.

Guideline or Requirement?

Although called a "guide," life in Tech Pubs will be a lot easier if the style guide and its rules are made mandatory. It's easier to follow the rules than to break them, and since nothing in the style guide should be incorrect, there should be no objection to following it. (Be ready to challenge the rules if they negatively affect clarity or any of the other important best practices, though!)

Following the rules makes peer editing a lot easier. An editor can mark up instances where a writer breaks the style guide, but it's unfair to the editor's time if a writer never intended to follow the style guide.

In some organizations, a style guide is often meant to be just that—a guide, not a book of hard and fast rules. Make a decision within your department as to whether your style guide is to be mandatory or just a guideline.

The Classics Are Always in Style

Besides the style guide created just for your company or department, you'll want a backup to use in situations that are not covered in your guide. There are several that are considered standard additions to a tech writer's library. Whether your products run on Windows, UNIX, Mac, cross-platform, or the Web, any of the following is a good choice to designate as a backup to your department style guide.

- *Chicago Manual of Style*
- *Microsoft Manual of Style*
- *Read Me First! A Style Guide for the Computer Industry*
- *Apple Publications Style Guide*

There are other published style guides tailored for writing in different fields. *U.S. Government Printing Office Style Manual*, for example, is an invaluable aid if you work for the US government or a company that does business with the government. Appendix B, "For Your Bookshelf," lists some other style guides you can consider. However, I recommend you use only one as your department's referenced style guide.

If you specialize or plan to specialize in a particular sector of the computer industry, there are special guides that apply. If an online search doesn't turn anything up, ask someone who writes in the field you're interested in.

Creating a Style Guide

Before you start creating your own style guide, see if one exists for your company. Often, Product Marketing is already working from a style guide, and you should definitely use it if one exists.

If your company doesn't have a style guide and you have taken it on yourself to work on one, I suggest you choose one of those listed above as a starting point. There's no point in reinventing the wheel. Those books have been in existence for years, so why start today from square one?

However, there will always be situations specific to your organization that are not covered in the classic style guides. Because of that, you'll also need a style guide targeted to your company or department. It will be your department's first line of reference.

Style guide issues are often determined by the preferences of the first person who creates the style guide. Let that be you if you want your preferences to rule.

Building Your Own

As you start to write or update a documentation project, make a note whenever you have a question or see a possible inconsistency. If you see that the writers in your department tend to do something inconsistently, make a note of that. Make notes of how you spell or capitalize a word, how you format elements such as bullet lists, and how you punctuate. Make a decision on what the usage should be, either by deferring to the published style guide you've chosen as an arbiter, by taking a vote, or whatever method you choose. Once the decision is made, make sure it is in your collection of style issues.Your collection should include common and frequently questioned issues, as well as issues that have many possible solutions.

There are no rights and wrongs in a style guide— only the way your department chooses to do it. Consistency is the key here. One company might choose to use title capitalization on their figure and table titles while another company uses only initial capitals. Neither is "wrong."

If you don't like the way your organization does it, don't waste too much time trying to subvert the style guide. It's as easy to follow it as not, and before long, you'll find its guidelines becoming second nature.

Ask the Product Marketing group for a list of all of the company's products and refer to that list for trademark usage, capitalization, and spelling. Talk to your the Legal department to come to an agreement about how to trademark items—do you put the trademark after each use of the name, or once at the beginning of each section? Or do you put a disclaimer at the front of the document and no trademark symbols at all?

Create the style guide in a format that can be easily updated and made available to the team, like a wiki. A wiki makes it easy to add content as you think of it, and when something is easy to do, it's more likely you will do it consistently.

How Do You Know Which Style Is Right?

Your teammates may argue with you about their own style preferences. Individual writers have their own way of doing things, and if there are no current guidelines, they're each used to doing things their own way. How do you know which style is "right?"

There is often no right or wrong. In the end, many style issues are only expressions of opinion. If you and your colleagues have differing opinions, try one of the following resolutions:

- Choose one of the published guidelines as your arbiter and agree to use its version of the disputed style.
- Take a vote and agree to go with the majority.
- Ask your manager to make a final decision.

We'll never have a divine decree to settle whether "email" or "e-mail" is correct, or if all bullet lists should be preceded by a colon. In the end, the important thing is to pick one style and go with it. Consistency is more important than your personal preference, so even if you feel really strongly that tables and figures should use title capitalization and variables should be indicated with angle brackets, while the rest of your team thinks something else, at some point, you've got to let go and be willing to compromise.

Getting Started in Style

If you are creating your own style guide and are not sure where to start, here is a sample template of the most common issues a writer will want to resolve. As you fill out these sections, you'll notice that some entries have information that overlaps into other entries. Create a cross-reference to send the user to the related section, by adding "See also <Section Name>" at the end of the section.

I've added my own recommendations in a few of the places; if you have no preference of your own, this can save you time.

In fiction, the language and the senses it evokes are important, whereas in technical writing, the *content*, and the information it conveys, are important. In technical documentation, the language should never draw attention to itself. The moment a reader of technical documentation stops to puzzle over an unusual use of a word or phrase, you have lost that reader.

Abbreviations

A lot of issues can be covered in the "Abbreviations" section. I've listed a few of them here, but you will probably think of more when you create your guide.

Indicate whether or not you allow contractions. Some companies do not use them at all; others do, especially for end-user web content.

This is a good place to tell writers to avoid using Latin abbreviations such as *i.e, e.g., et al, etc.*, and others.

Here you might also include a list of all abbreviations commonly used within your company's documentation, or refer users to an online glossary of technical abbreviations that everyone is to follow. Think of how you abbreviate things like measurements and time (am, a.m., or AM?).Cross-reference to Acronyms and Product Names.

Accepted Terms

This section can contain an alphabetical listing of accepted words such as email vs. e-mail, prepay vs. pre-pay, sign in vs. log in, dropdown vs. drop-down, and others. Include every example you can of words or terms that have alternative choices. Cross-reference to Capitalization, Hyphenation, and Word Usage.

Make sure the style guide is written to follow its own rules. Users find it highly confusing (not to mention irritating) when the convention states to do something one way but the guidelines are not followed. Some people will do what the guide says, others will do what the guide shows. Others will complain, and rightfully so.

Acronyms

Although technically there is a difference between an acronym (a pronounceable word formed by the initial letters of a name, like "LAN") and an abbreviation (a nonword formed by the first letters of a name), most people don't know the difference and it is not relevant to style usage. What the Acronyms section of a style guide typically explains is how to first refer to an acronym versus the full name.

My recommendation:

> The first time you use an acronym, spell out the full name of the feature, product name, or concept and place the acronym in parentheses after it. After that, use just the acronym. If the acronym is much more common than the spelled-out term, you can put the acronym first

In this section, you should also list all the official acronyms for your company's products. Include those that are not permitted as well, such as internal, outdated, and misspelled versions. There are always a lot of outdated and unofficial names in use internally, and it's important to call these out to prevent writers from using them in the documentation. Cross-reference to Abbreviations and Product Names.

Bullet Lists

There are so many ways to format bullet lists, this section becomes very important. Let your users know about lead-ins to bullet lists, whether or not to capitalize list items, and how to punctuate the list items.

My recommendations:

- Precede all lists with an introductory clause that ends in a colon. This can be a complete or partial sentence.
- Start all list items with a capital letter unless the word is a product name that begins with a lowercase letter.
- Do not put punctuation at the end of a list item if it is a fragment or if it completes the sentence started by the introductory clause.
- End list items that are complete sentences with a period..

Capitalization

This section often lists words and terms that are capitalized, such as product names, department names ("Product Management" vs. "product management"), and initial letters in bullet lists.

It also includes capitalization rules for headings, table titles, and figure titles. This can range from all uppercase to title case to sentence case. *Title case* involves capitalizing the first and last words of a title, and all nouns, pronouns, adjectives, verbs, adverbs, and subordinating conjunctions (such as *if, because, as, that*). Don't capitalize articles (*a, an, the*), coordinating conjunctions (words *and, but, or*), and prepositions.

Different style guides have different rules. Some say to capitalize prepositions and conjunctions of four or more letters, some say to capitalize prepositions and conjunctions of five or more letters. Pick the style you like and follow it.

Sentence case is the easiest case to use—with this method, you capitalize the first word and all proper nouns within the title and the rest of the words are lowercase. This is more commonly used in countries other than the United States, but if you dare to try it, I think you'll like it. (No more worrying about what words to capitalize within a heading.) Cross-reference to Product Names and Bullet Lists.

Clicking and Selecting

Writers should use consistent terminology and typography when directing users to select or click a link or button on a software application.

My recommendations:

- *Click* a button or link. Don't repeat the word "button" or "link;" that is, say "Click **Submit**," instead of "Click the **Submit** button."
- *Select* from a menu or toolbar. For example: "Select **Save** from the File menu."
- Use bold to indicate any link or item that can be clicked or selected.

Contractions

Here you can specify whether or not you allow contractions to be used, or simply cross-reference by indicating "See Abbreviations."

Hyphens

Here, you can list words that are hyphenated. (Or not hyphenated, as the case may be.) Cross-reference to Accepted Terms.

Numbers

Indicate when numbers are spelled out and when the numerical form is used.

My recommendation:

> Spell out numbers from zero to nine and use numerals for numbers greater than nine. The exception to this is when a number is used as part of a name (EnterPrize Cloud Storage version 10.1, for example), or when a number is one million or greater. For large numbers of this type, use the numeric value and then the word—for example, 10 million or 2.5 billion.

Product Names

List every one of your company's product names with correct trademarks, spellings, and capitalizations, along with allowed and non-allowed usages. This is where you tell writers that they are not permitted to refer to EnterPrize Cloud Storage by its internal nickname, "ECS," and that the "P" must be capitalized with no space in the word.

Cross-reference to Trademarks.

If your typeface has a zero that looks too much like an uppercase "O," you might create a rule in your style guide that "zero" should be spelled out in parentheses after the numeral, as in the following example: **0 (zero).**

Punctuation

You don't have to include every punctuation rule here; just concentrate on situations where there is more than one correct way to do something. Here you include items such as:

- Always use serial commas.
- When forming a plural of an abbreviation, don't add an apostrophe.
  Correct: **URLs**.
  Incorrect: **URL's**.
- Use a colon to precede a list.

Trademarks

In this section, explain how and when trademarks, registered trademarks, and service marks are used. Typically, these marks are used the first time the marked name appears in the content. After that, the name is referenced without the mark. If your company has a department responsible for this type of thing, discuss it with them. If not, see how other companies like yours handle trademarks and do something similar.

Include a list of all of your company's product names that are trademarked or service-marked, with the correct mark after them. If the documentation frequently refers to other companies' products, include their trademarks as well.

Cross-reference to Product Names.

Typographical Conventions

You should understand your department's typographical conventions so you know when to apply the correct character tags, style, markup, or formatting to a word or term within your text. Most of the methods you use to produce text will include markup of some sort. Instead of selecting a word and making the characters bold, you apply a designated tag to the word, which will cause the correct style to be applied in the output. Your style guide should include a list of the accepted styles used within your documentation, something like the following:

Item	Character style	Rendering
Buttons	<Button>	Bold
Variables	<Variable>	Italic
Emphasis	<Emphasis>	Italic
User command line input	<UserInput>	Monospaced bold

However, if you are working in Microsoft Word or a tool like it, you will have to apply character formatting to individual words or terms. (Word's styles work at the paragraph level, not at the individual word level.) And because the style guide is often shared across the company, where individual departments using their own tools will create their own methods of applying the style, I think it's important to include a description of what the typographical conventions are. (For the end user, the typographical conventions look like the table shown on page 230.)

To that end, I still like to include the rules for typography, not only to tell a writer or the person developing the style markup what to do, but also what not to do.

My recommendations:

Typographical conventions

Convention	Description
Bold	Use bold type to indicate something that a user clicks, selects, or otherwise acts on. Examples: Click **Next**. Select **Settings** from the Edit menu. Do not use bold for names of menus, screens, web pages, dialogs, or windows.
Italic	Use italics for variables, document titles, and terms that require emphasis. Examples: Enter *host*, where host is the IP address of the web server. See *EnterPrize Cloud Storage Administrator Guide* for more information. Make sure all files have been transferred *before* exiting.
`Monospaced text`	Use monospaced text (fixed font) for code samples, system output, and path names.
`Monospaced bold`	Use monospaced bold text for user input on the command line.

Word Usage

In this section you go a bit further than you did in the Accepted Terms section by including what *not* to use as well as what to use. You can list all words that are to be used in your documentation, with a two-column table with the correct usage on the left and incorrect usage on the right. For example:

Use this	Instead of this	Comments
email	Email, e-mail, E-mail	
sign in *when it is a verb*	login, log-in, logon, log-on, sign-in signin	"sign in" is a verb For example: Sign in by entering your username and password.
sign-in *when it is an adjective*	sign in	"sign-in" is an adjective. For example: The sign-in page...
website	Web site, web site, Website	

Finishing in Style

At the beginning of the guide, tell your users what published style guide they should consult if their issue cannot be found in your department guide. If the published guide is available online, provide the URL.

The *Chicago Manual of Style* is one of the top choices of writers everywhere. Consider an online subscription, available at **chicagomanualofstyle.org**. Subscribing gives you access to all the newest information, online tools, and user forums.

Once you have a solid draft of the style guide, plan to share it with the other members of the team. They are sure to have their own ideas of what needs to go into the guide. It can take several rounds of discussion before it is finalized, but the other team members are much more likely to accept the guidelines if they have a say in them.

When the guide has enough information in it, post a PDF or HTML version of it on your department's website and make sure you, and everyone on the team, has access to it to update regularly.

Share the style guide with the rest of the company, too. Any department that does writing will be happy to have a written style guide to follow.

Procedure Guides

Style guides often contain procedural information such as instructions about how to do things within the Technical Publications department. Procedure guidelines are very useful in a new organization or one that has a lot of contractor help or turnover. If you are a lone writer and want to leave a legacy, or if enough people in your department ask frequently how to do things, it's worth it to assemble a procedures guide.

Procedures include anything and everything—such information as how to generate online help, what size and format a screenshot should be saved as, how to structure directories in a version control system, and how to build books out of templates. Write them up as you think of them. You, and the other writers on the team, will be glad you did.

Chapter 19

Front and Back Matter: Or Do They?

How to give your documentation meaningful entrances and exits.

What's in this chapter

- What you'll find at the beginning and end of a book
- Why indexes are still needed
- Making an index that's better than a search

Your deliverables might be anything from a 140-character Twitter message to a 1,000-page PDF guide. In all cases, there are final touches that show that these are finished works. If the deliverable is a set of help topics, you want a splash page (that is, a page users land on when they first come to a site) or a home page that ties it all together, with some introductory material. If you are writing a manual, you want to present it in book form, and that usually includes what is called front matter and back matter.

With less dependence on the book form in technical documentation, you might think that the components described in this chapter are old-fashioned and no longer apply. Not so. There are still users who rely on the book format. And for a book to be finished, it must have certain parts.

This chapter talks about some of those parts and how to create them.

Let's Be Upfront

The first few pages, or front matter, of a book usually consist of:

- Title page
- Copyright page that includes information such as trademarks and acknowledgements
- Table of contents
- Introduction

Title Page

The title page should include a few key parts: the title of the document and the name and logo of your company. Depending on your company's style, there are other components that can be appropriate on a title page, such as version number, date, and confidentiality notice.

It's important to make sure your title page meets corporate branding standards. Use imagery, typography, and color schemes provided by Marketing. If you don't feel comfortable designing a page that adheres to branding standards, ask someone in the Marketing department if they can provide you with a layout.

Copyright Page

The copyright statement, or notice, normally appears on the inside front cover (or page 2) of a printed or PDF document. In online content, a copyright statement may appear on every page the user sees. The copyright is very important because it establishes ownership of the document's contents.

The accepted format of a copyright is the word "Copyright" followed by the copyright symbol (©) followed by the year, the name of the copyright owner (that would be your company, not you!) and the words "All rights reserved." It might also include a disclaimer spelling out what the copyright notice means. In a technical document, this will be something along the lines of "No part of this publication may be reproduced, stored in or introduced into a retrieval system, or transmitted, in any form or by any means (electronic, mechanical, photographic, audio, or otherwise) without prior written permission of [Company Name]."

If all of that sounds like complicated legalese, that's because it is. Do not attempt to craft this statement on your own. Consult your organization's legal counsel and ask them for the correct wording. Check in with them regularly to make sure it has not changed—you might be surprised at how often this wording changes as the company refines its message, gets new legal people, and reviews content.

One decision that affects you is how to handle revisions and updates made after the initial copyright year. For example, let's say you write, from scratch, the *EnterPrize Cloud Storage User Guide* in 2012. You include a copyright statement that reads, "Copyright © 2012, EPCS Networks Incorporated. All rights reserved."

In 2013, you do major updates to that document for a new release. Companies frequently expect the statement to be changed to read "Copyright © 2012-2013, EPCS Networks Incorporated. All rights reserved." What if you change the title of the document in 2013 so it's now called *EnterPrize Cloud Storage 2.0 User Guide* but it is based on the same original content? You need to decide if this is a new copyright date, or you should indicate the date as 2012-2013.

The *Microsoft Manual of Style* recommends that if 85 percent or more of a document is new, it gets its own brand-new standalone copyright date; if less than 85 percent is new, a date range is used based on the year the document was first published.

But I'm not a lawyer, and this advice is only a recommendation to help you know what questions to ask. Legal departments don't always think about technical documentation, so it becomes your job to make sure the right questions are asked and the legalities are correctly carried out.

Table of Contents

The table of contents is an important part of your front matter. Even though a PDF can have its own table of contents (called Bookmarks), a table of contents for a book is important for users who like to print out and read a manual. If you are creating online help, a table of contents for help is important to let users see at a glance what is in the help and where to click to find the information they need.

Although many users prefer search techniques, and want to type a word in a search box, other users appreciate a table of contents. Do both—it is another part of what gives your document its final, professional, polished look.

Indents, font sizes, and bold fonts should help cue the reader to the level of the heading. Luckily for us, all publishing tools can automatically generate a table of contents, so set up your tool so it creates a table of contents that you like.

I recommend that a table of contents be no more than three levels deep. A table of contents that is too long is not terribly useful; a reader can get lost paging through it. A good rule of thumb is to include chapter titles, and heading levels one and two. I used a three-level table of contents in *The Insider's Guide*, but because my top-level heading is a Part heading and this book has many chapters and headings, I did not include two levels of headings within the chapters .

Once you generate a table of contents, try to avoid tweaking it afterward by making format changes to the source files. If the table of contents breaks a heading in a way you don't like, rewrite the heading in the original chapter so it looks better in its final form. Otherwise, you have to remember to modify the output each time, and guaranteed, the one time you forget to do it will be when the document is going out for its final trip.

Tables of contents are good tools for reviewing your document's structure. Have you been consistent from one chapter to the next? If you have a Heading 1 called "What's in this Chapter" in every section but one, you can see that gap easily. If you have a Heading 2 without a Heading 1 above it, you can see that too. If your chapter title style is to use gerunds, you can see immediately if you forgot to do so in some of your sections. Generate your table of contents regularly to make sure you are structuring your document correctly. (You can even generate a table of contents with all levels in it to check for problems like this, and remove the extra levels from your final table of contents output before you go to final production.)

There was a scene in the 1997 movie *Air Force One* in which the U.S. president, played by Harrison Ford, is holed up in the baggage compartment of the presidential plane while terrorists hold the plane's other occupants captive. Luckily for him, he finds a cell phone in someone's bag...but he doesn't know how to use it. Frantically, he pages through the accompanying set of instructions while the lives of his loved ones and staff hang in the balance.

He could have used a good table of contents and an index!

What about additional tables of contents, such as the separate list of tables and figures you see in some documents? I don't believe that these are very useful in most documentation, and they end up looking like a way to pad page count. Tables in a document are usually created because they are a useful means of presenting information, not because they are so important that they deserve their own contents list. A good table of contents should help users find the figure or table they're looking for by having clear titles. If a table or figure is so important that it should be called out somewhere, mention it in the index.

However, if a separate list makes sense and provides helpful information to your customers, then by all means, provide it. If the document is a catalog, for example, it might be valuable to have a list of all orderable items in numerical order by part numbers. In a command reference, you might want a list of all commands in alphabetical order.

Introduction

"Hello, nice to meet you, my name is..." That's what your introduction needs to convey to the reader, and it needs to do it in a way that is succinct. After all, you don't like meeting someone who rambles on garrulously, do you?

Here are the points your introduction might cover:

- A brief description of the product about which the document is written
- An overview of the document's purpose (How to use the product? How to install it? Administer it?)
- A high-level overview of the information the document contains
- Who the document is written for and what level of expertise they are expected to have
- Information that applies to the entire document or is required before reading the rest of the document, such as a product matrix, a list of prerequisite software, or a conceptual overview of how the product works.
- A list of available related documentation

Keep the introduction fairly short and make sure it contains useful information. If it's not useful, it doesn't need to exist in your manual at all.

Other Front Matter

You don't want to clutter up the beginning of your documentation with too many pages (it gives the readers more to skip over) but there are a few other items that sometimes appear in the front of a book.

Typographical conventions

Some documents—but not many these days—include a typographical conventions section, where the user learns what typefaces are used to convey user input, variables, code, and other special information.

We talked about typographical conventions in Chapter 18, "The Always-in-Style Guide." If you are consistent in your use of typography, the users should understand the meaning, the same way they understand which lines in a book are headings and which are body text.

If you feel there is any possibility of users not understanding the conventions, add a typographical conventions table to the beginning of your documentation.

Your typographical conventions should look something like the following:

Typographical conventions table

Convention	Description
Bold	Bold type indicates something that you click, select, or otherwise act on. Examples: Click **Next**. Select **Settings** from the Edit menu.
Italic	Italic type indicates a variable.
`Monospaced text`	Monospaced text (fixed font) indicates code samples, system output, and path names.
`Monospaced bold`	Monospaced bold text indicates user command line input.

Warnings and cautions

The front matter sometimes contains descriptions of cautions and warnings, as well as the iconography that can accompany them. If your document contains cautions or warnings, these descriptions can be helpful, since it's important for the user to understand and recognize them, for example:.

! A caution means there is possible damage to equipment.

A warning means there is possible danger to a person.

Revision information

If the document is a revision or update of a previous document version, it is helpful to include information at the beginning describing the changes that were made. This enables users familiar with the older document to see at a glance what the changes are to the new one.

About Back Matter

Back matter is material that is found, well, at the back of a book. Back matter includes sections that are not part of a linear progression of information and are therefore not expected to be read by all users of the book. It might also include one or more appendix sections, a glossary of terms, and an index.

Appendixes

An appendix is a collection of supplemental information at the end of a book. It is typically made up of reference material that does not belong in the main body of the book. This could be more detailed, in-depth information that is not necessary for using the product but is interesting to some users, or specialized information for optional features that not all users have.

To indicate their difference, appendixes usually number their titles with letters instead of the numerals used by chapters.

Glossary

A glossary is a list of specialized words and their definitions. The words should all be relevant to the documentation's subject matter and should be listed in alphabetical order.

Glossaries don't always belong in the back of a document. Depending on your business, your glossary might have so many entries that it requires its own separate document. If so, make sure it is available to all customers, ideally in a searchable online form, so they can look up any term about which they are unsure.

Index

An index is still a very important part of a document. (And by the way, indexes add value to online documentation as well as print documentation. Many help systems include indexes.) Why do we need indexes, you might ask, when users can search for any term they want? Well, there are several reasons.

For one thing, the document might be printed out rather than accessed online. Users have no way to do a word search on a printout, so the index helps them find information. In addition, an index provides context. It doesn't simply list all words and phrases found in a documentation, but rather provides listings of terms and ideas that are reworded versions of what appear in the document. The index also provides lookup information at a much more granular level than does the table of contents.

If your document is 20 pages or more, you should include an index, although you are certainly free to create indexes in shorter documents. However, with PDF and search engines and good tables of contents, an index can often be omitted from shorter documents with no great harm.

Plan on your index being one column of entries for every 10-20 document pages. (An index page is usually a two-column layout, although it can be three.) Your index might be shorter or longer depending on the complexity of the material.

Manual or Automatic?

While indexing software exists, it's not necessary. If you build books with Adobe FrameMaker, Microsoft Word, or similar desktop publishing tools or word processing programs, or use a standard help authoring tools, these programs have their own built-in indexing functionality., and the major XML markup languages have elements for index markers. You enter a *marker* (an indicator that the marked word or term belongs in the index) into the source file, and when you generate the index, the marked terms are compiled in alphabetical order into the index.

Indexing software tends to gather keywords and is not able to index anything based on concepts. It can be a helpful start, but make sure you use it only as a starting point and modify and improve your index entries using the tips in this chapter. I recommend doing all indexing yourself without the help of indexing software, by going through your documentation (who knows it better than you? Certainly not a piece of software) and marking up the terms yourself. If you do feel you must have the help of software, do a web search for "indexing software" and try out some of the ones you find.

You might also look for software that helps you do better manual indexing by helping with some of the clunkier aspects of your documentation tools. For example, Silicon Prairie Software produces a plug-in for FrameMaker that helps with conditional text in index entries, formatting page ranges, and more. If you find that some aspects of creating an index slow you down, look for plug-ins of this type.

What an index looks like

In its simplest form, an index is an alphabetically ordered list of terms (topics) with locators that point the user to the place the term can be found in the body of the work. The locator might work with a hyperlink that takes the user directly to the place, or it might display a page number that the user goes to. The locator might also refer the user to a different topic to find the information (with a see reference), or to find additional information (with a "see also" reference). A topic can have one or two subtopics.

Each entry should be descriptive enough to let the user know what to expect when she arrives at that location. For example, instead of simply sending her to every instance of the word "file," you send her to a location that provides specific information about files, and you tell her what that information is.

The *see* locator is useful for acronyms and abbreviations, as in an entry like this:

VPN, *see* Virtual Private Network

As a rule, however, you might want to use *see* only when the primary topic has many entries. If the topic you're sending the user to has only a single locator, you may as well do the user a favor and cite the page number in both places instead of sending the user to a topic that provides nothing new.

That's it. The output will look something like this:

- files
  - deleting 18-19
  - editing 24
  - excluding 44-45
  - recovering older versions of 49-52
  - selecting for sync 27, 28, 29
  - sharing 16-18, 26, 33-34
  - syncing between computers 27-28
  - uploading 12-14, 26, 29, 32-34, 51

In the above example, "files" is the topic, a first-level index entry. Each of its subtopics, or second-level index entries, include locators that show their different pages and page ranges. Although you can go down many levels, I would suggest you stick with a maximum of two levels until you gain expertise in indexing. An index that provides three levels would look like this:

- files
  - deleting 18-19
  - editing 24
  - excluding 44-45
    - from sync 44
    - from backup 45
  - recovering older versions of 49-52
  - selecting for sync 27, 28, 29
  - sharing 16-18, 26, 33-34
  - syncing
    - between computers 27-28
    - settings for 27
  - uploading 12-14, 26, 29, 32-34, 51

As you can see, you can be quite descriptive when you go three levels deep. In the above example, the first level is "files," and the second level is "excluding" and "syncing." The subtopics below each of those provide more detail.

What makes a good index

An index should be better than a word search. An index should not only show the user the locations of a term, but also it should include other terms the user might think of that don't literally appear in the document.

For example, suppose the user wants to know how to delete certain files that are stored with HomePrize Cloud Storage. If she does a word search on the word "file," she'll get thousands of hits, with many of them being on the same page. She might try searching on the word "delete" and discover there are also too many hits to easily sort through, and then try "deleting" and find that it's too limited.

A good indexer anticipates this kind of search by thinking of every term and concept a user might come up with and making sure they appear in the index.

So let's expand the previous index sample a bit. In the following sample, I've added a number of cross-references and alternate entries for each term.

- deleting
  - files 18-19, 21
  - folders 18-19, 21-22
  - users 57
  - versions 48
- files
  - deleting 18-19, 21
  - editing 24
  - excluding 44-45
  - recovering older versions of 49-52
  - selecting for sync 27, 28, 29
  - sharing 16-18, 26, 33-34
  - syncing between computers 27-28
  - uploading 12-14, 26, 29, 32-34, 51
  - *see also* versions, folders
- folders
  - deleting 18-19, 21-22
  - excluding 44-46
  - selecting for sync 28-29
  - sharing 16-18, 26, 33-34
  - syncing between computers 26, 28-31
  - uploading 29, 33-34, 51,52
  - *see also* files
- removing files, *see* deleting files
- storage limit
  - about 4-5, 9
  - upgrading to higher 5-6, 10, 58
  - exceeding 5, 9-10
- storage quota, *see* storage limit
- syncing
  - about 3, 4-7
  - files 4, 24-29
    - between computers 27-28
    - to File Manager 24, 26
  - folders 24-29
    - between computers 28-31
    - to File Manager 24, 26-27
  - selecting items for 27, 28, 29
- quota, *see* storage limit
- uploading files 12-14, 26, 29, 32-34, 51
- users 56-58
- versions, 12, 46-54
  - about 12, 46
  - deleting 48
  - managing multiple 47, 51-52
  - recovering 48-49
  - undeleting 50

Not all users will look under "files" to understand how to delete files. They might, instead, look under "deleting." And they might also look under "removing" or any of a number of other types of terms. The expanded example includes all you really need to know about an index:

- It has a number of ways for a user to look up one concept. For example, a user can look up "syncing" by looking under "files, syncing"; under "folders, syncing"; or under the single entry "syncing." For each of these entries, a user can see all the pages he might want to go to. In the same way, a user can find "deleting files" under "deleting" or under "files."

- As you look through the topics, you'll see that they are repeated and related to each other throughout the document. For example, "deleting" has its own entry but you can also find "deleting" under" files," "folders," and "versions."
- It uses the "see "locator to point the user consistently to a topic where the user can find information about topics that have similar meanings. In this example, "storage quota" and "quota" both appear in the index. Both of these entries point to another entry, "storage limit," which means more or less the same thing. Instead of repeating the subtopics and page numbers for every term that has a similar meaning, "storage quota" and "quota" do not include a page number. They point the user to "storage limit," which has more information and subtopics under it.
- It uses the "see also" locator to point the users to a term where they can see more details about a related topic. So a user who is looking in the "files" topic might also be interested in going to the "versions" topic where she can find locators for "file versions."

COFFEE BREAK

You might have heard about or seen "joke" entries in indexes such as the famous entries for "loop, endless—*see* endless loop" and "endless loop—*see* loop, endless." Don't be one of the many tech writers who create similar endless loops in their indexes without intending to be humorous.

So that's how you start working on an index. And "start" is the operative word here. You would actually include many more second-level entries for these topics, and many other first-level entries based on all of the ways users might think.

When Writers Hate to Index

It's not uncommon to hear tech writers groan when told they have to index their work. Many writers hate it. They don't like to go back after documentation is finished and figure out what should be indexed. They don't like to index while they write, stopping to add index entries. And they don't like to remember to add new index entries when updating documentation. It can be tedious and time-consuming and it has to be generated several times before it's right.

And many writers don't understand what makes a good index in the first place. They confuse the purpose of the index with the table of contents or glossary. These writers know there are professional indexers and wonder why they aren't being hired to do this work.

Hiring a professional indexer is not an option for most Tech Pubs departments; there's rarely the time or money for it. No, indexing is usually the responsibility of the technical writer.

A *topic* (also called a *subject*) is the word, phrase, or abbreviation listed in alphabetical order in the index. Together with the reference, topics make up the index entry. A *reference* (also called *locator*) is the part of the index entry that takes the user to a place that answers his question. It's a page number or range of page numbers, or a pointer to another subject, typically indicated with "see topic xyz."

Practical indexing tips

There are a few tricks you can use to make it easier to create an index that is more likely to be right the first time. Making consistent index entries lessens the number of "repair cycles" needed to tweak the result after each generation.

- Use lowercase letters for every word except proper nouns. It's easy to remember and you avoid having one entry that says "Managing Users," another that reads "Managing users" and still another that is "managing users."
- Use plurals for all nouns, except when the term in question is one of a kind. For example, a topic would be "servers," not "server."
- Use gerunds (-ing endings) for all verbs, such as "installing" or "adding."

You might wonder when the best time is to create an index—whether you should add the markers as you go along, or finish the document and then do the indexing. There's no best way—it's a matter of personal preference. Some people can race through a completed document and produce a finished index in no time. Others find it tedious to add markers all at once and prefer to do it gradually as they write.

No dead ends

The most important thing to remember is that you need to keep your index up to date. When you edit or delete content, be careful not to delete important index markers. When you copy material from someone else's document, make sure you don't keep incorrect entries. When you move text from one part of a document to another, make sure you carry the index entries along with the content. And when you do a revision or update, do not forget to add new index markers for the new content.

It takes time and practice to do good indexing, so I wouldn't expect you to be an expert right away after reading this short section. But a good index is like a work of art, so practice away and you'll be on your way to mastery.

Chapter 20

Design and Layout

Knowledge and advice for design novices, and tips on how to create a template.

What's in this chapter

- Adding good design to your tool kit
- Why the grid is important
- Why less is more
- Putting it all together

You're a writer, not an artist. You've got no training in graphics, design, or art-anything. When you were in grade school, your teacher took your crayons away. But now you're a tech writer and you're expected to come up with not only the content of the documentation, but also the layout and templates for both PDF manuals and help.

Well, as you must have figured out by now, a technical writer wears many hats, and the artist's beret is sometimes one of them.

Whose Job Is It, Anyway?

I suppose it's no surprise that today's tech writer is often expected to own design and layout in addition to writing. Publishing, illustration, and photo-retouch programs have made it possible for us to do work that used to take a staff of layout and paste-up artists. If that's not your background, then it can be difficult to create a document with a pleasing and professional layout.

Take What You Can Get

Luckily for you, you don't have to figure out everything yourself. Desktop publishing programs and word processing programs come with templates all ready for you to use to create your department's documentation. If you are designing online help or web-based content, many web development applications also come with templates.

Take a look at what other companies have done with their technical documentation. Not for stealing, but for learning. Do an online search and look at as many documents as you can, stopping to think about what works and what doesn't, what typefaces are nice to read and how the text sits on the page.

Obviously I would not suggest you duplicate another company's look and feel. First, your company wants its own look, and second, you don't want to raise issues of copyright violation. But you certainly can borrow ideas such as type size and style, column width, and placement of headings on the page.

So use the templates that come with your applications, and gather inspiration from people who have been there and done that already. With some minor modifications, you can make these designs your own.

Get Help Where Help Is Needed Most

Illustrations can greatly enhance documentation. They let a user see concepts that are difficult to describe. But all illustrations are not alike. Although I know you can learn the basics of layout, typography, and template creation, as I said earlier, technical illustration is a whole different animal and one I would not suggest you volunteer to take on.

Sure, you can and should do charts and graphs and simple line drawings. And there is a lot of good clip art available too, when you need it. But complicated technical illustrations, including those that show exploded views, cutaways, and prototypes of products that don't exist yet, take a lot of skill and training that are rarely found in a tech writer's bag of tricks. If you find yourself needing these types of visual aids, seek out the help of a technical illustrator.

Enhancing Usability with Visual Elements

Visual effectiveness is an important factor in usability. For the tech writer, visual effectiveness is achieved through:

- Illustrations that create clear pictures for the user
- Good layout and design

A poor design can create confusion when instructions are buried in blocks of text, warnings and cautions are hard to find, and text is just too small. A document that is hard to read either won't be read or will be read and misunderstood.

A good design goes a long way toward making a good document. Adding *white space* (negative space, which is the space between elements) not only can help create a sophisticated and interesting design, but also has been proven to increase readability. Have you ever had to read a big, closely spaced block of text that was very wide and the type size very small? The same block, broken into several paragraphs with more space between the lines, is much more readable as well as easier on the eyes.

When you think about good design, be aware that too much design can be as much of a problem as too little design. If you are producing marketing or sales documentation, you may want to be a little splashier than if you are producing straight technical documentation. For technical documentation, you should be thinking about one thing: ease of use. If your choices are between "boring" and "exciting," but "exciting" is splashy, distracting graphic design that is hard to read, go with boring. Exciting and innovative design is a great thing, but not always best for technical documentation.

That's not because our users don't appreciate color, jazz, and eye appeal. The reason is simpler than that. To convey information clearly, a document should not contain any distraction or elements that make it hard to read. The reason for distracting elements is usually commercial—the designer is trying to get the reader to notice an ad or a text block. Since most technical documentation doesn't have to worry about advertising, you don't need to lure your readers away from your content.

But don't put your users to sleep! You can still provide a clean, clear, attractive design for your users by keeping it simple.

Creating a Template

The template is where all of these design decisions come together. Whether you create books, help, web content, mobile device content, or any other type of content, a template—like a style guide—lets you create consistent documentation. Using a template also helps save time because you don't have to think about layout each time you build a new document, and you don't have to manually format each heading or paragraph.

Building a template is a lot of work, but it saves a lot of time later and helps your documentation look great. You'll need a template for each different page type: title, table of contents, main content, and index. If each chapter has a different kind of front page, or if you sometimes need landscape-oriented or other different types of pages, you'll need templates for those, too.

You will probably need a different template for each section type: chapter file, title page, table of contents, index, release notes, and online help topics. A documentation template contains all styles that are used in that particular section. For example, an index template might have styles defined for the index title, index letter, the different index levels, and the headers and footers on the page. A chapter template, on the other hand, needs styles defined for everything from body text to all heading levels. to figure titles to the same headers and footers, but it does not need the index styles.

Left/Right Book Pages

When you design a template for a book, you need to decide whether you want your book to have left/right pages or be generated as a single-sided document.

Left/right pages are designed for bound books, and the orientation of the pages is based on the page's inside (toward the center binding) and outside (toward the cut edge of the page). *The Insider's Guide to Technical Writing* follows that layout because it is bound in a cover.

Left/right book pages are usually designed to be a mirror image of each other (although they don't have to be). This can result in margins that are wide on the right side of a left-hand page and on the left side of a right-hand page, to allow extra white space in the center where the book is bound. The headers and footers are typically mirror images also, with page numbers typically on the outside edge.

You might also notice that the first page in each section or chapter of a book starts on a right-hand, or odd-numbered, page. To force this structure, all sections must end on a left page. If there is no content on the last left page, the publishing tool creates a blank page. Some companies have traditionally added "This page intentionally left blank" to an empty page, although that is no longer common and I wouldn't suggest you do the same.

Many technical documents use this standard book structure. The assumption is that a user will print the entire document, attach it at the upper left corner with a staple or clip, and flip through it as they would any book.

Single-Sided Pages

Single-sided pages all have the same orientation for headers and footers, and equal-sized margins on the left and right. There is no differentiation between the left page and right page, so therefore no need to add a blank page anywhere.

Writers who prefer this method believe that most users of PDF documentation read it on screen. If they do print, they often print only a section of the document. Blank pages at the end of sections waste paper and mirror-image headers and footers are unnecessary.

It's a matter of preference whether you choose a left/right page layout or a single-sided layout. They both work.

Size Matters

If you are designing for print manuals (by print, this will usually mean PDF), it is pretty certain your page size will be the same as standard printer paper size: 8 ½ x 11 inches in North America, and A4 (21 x 29.7 cm) in the rest of the world.

If you are designing a layout for a document that will actually be printed on paper, then you can choose—and *should* choose—a page size other than a full page size. Talk to your printer about standard sizes. You'll find that some sizes are less expensive than others, so it's good to know what your options are.

In fact, you *don't* want to make the page sizes smaller than a standard page size, because users need to be able to print them out on whatever paper they have in their printer. If you set up your PDF document for, say, 7 x 10 inches, instead of having a well-designed page, you would end up with too much white space on all sides.

This is what I think is the down side of the electronic documentation world—the fact that with the advent of self-printing, much of what we read is unbound and on too big a page size for really easy reading.

Your layout should look good on *both* an 8 ½ x 11-inch and an A4 sheet of paper...remember it might be printed out by people all over the world.

By now, you have some information about your document size and layout and are ready to start building the template. Let's assume that the page size is US Letter (8 ½ x 11 inches), but we know it needs to work equally well on an A4 size.

Page Structure

To design your page, I suggest you make a grid. (This works for web pages, too.) A *grid* is an invisible structure of horizontal and vertical lines that defines where elements go on your page. This helps you maintain consistent placement of all titles, text components, tables, figures, and margins. This consistent alignment creates a better design.

The templates that come with your publishing tool contain simple grids. Use them as a starting point or create your own, like the one below. Page 251 and 252 show pages based on this grid.

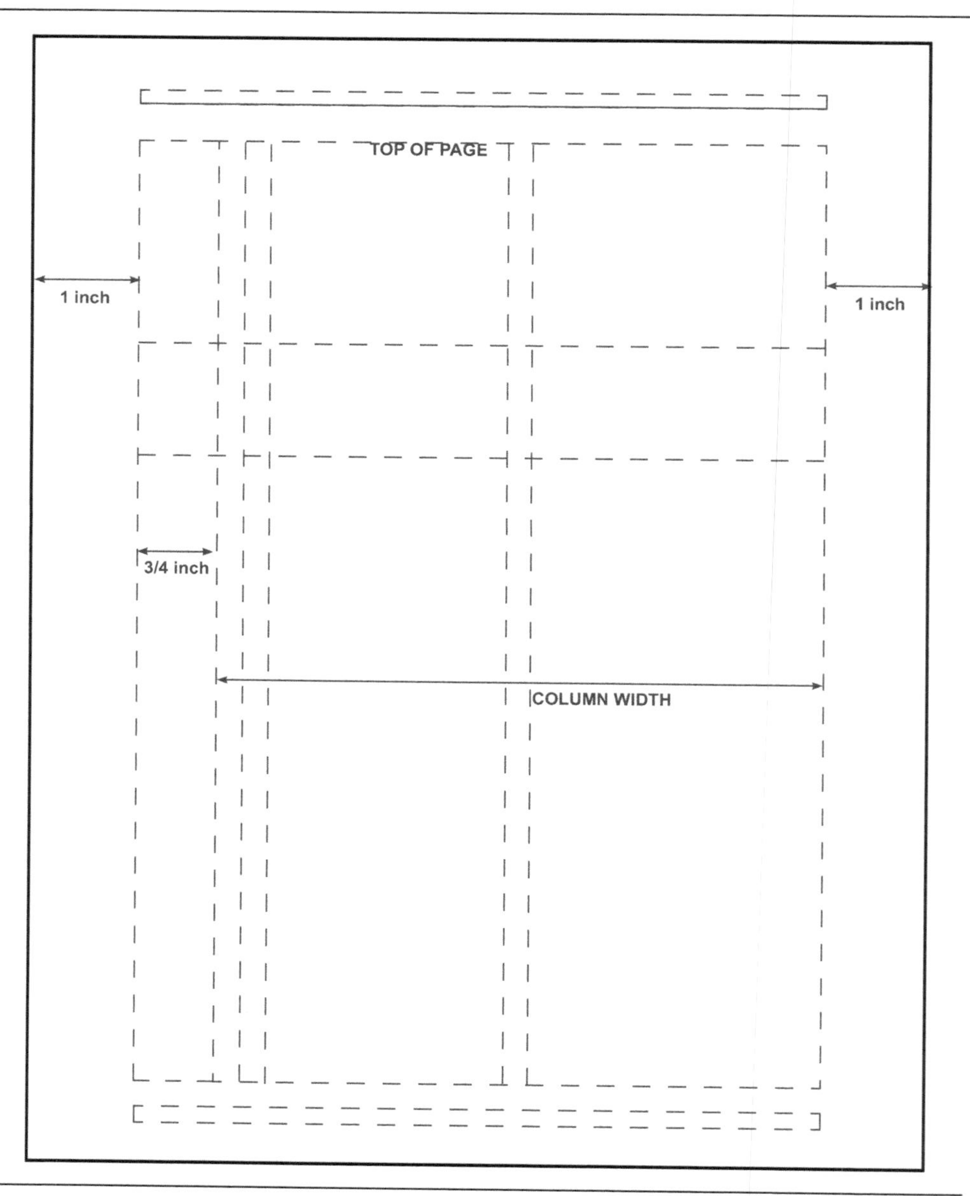

Grid for a standard US letter size page

Designing the Page

Whether your page is web-based or print-based, a design's readability depends on a number of factors:

- White space
- Line length (the width of a line of text)
- Type style
- Type size
- Leading (the vertical line space between one line of text and the next)

White Space—It's Not Always White

"White space" is a key part of any design. *White space* refers to the negative space on a page. This means that on a page printed on white paper, the white space is white. But on a web page with a black background, the white space is actually black!

Confusing as that might be, all it really means is that you should not cramp or crowd the elements on your page. Densely packed pages and lines of text that are too close together are hard to read and are not visually pleasing.

Do a quick scan over your documentation and make sure it doesn't contain long, unbroken blocks of text. If it does, add headings, bullet lists, figures, and tables to help break it up.

Line Length and Column Width

One way to ensure sufficient white space is to limit the line length; that is, the horizontal distance from the beginning of a line of type to the end.

There is no hard and fast rule for the best number of characters per line. You want something that is neither too short nor too long. If the lines are too short, your head has to move back and forth too often, and if they are too long, your eyes can lose their place.

Whether designing for print manuals or web content, a line length of 55 to 75 characters (including spaces) is generally considered to be easy to read. You'll see studies that say that 39 characters per line are best for readability, but that is a very unrealistic line length for any significant amount of documentation. After all, we're not trying to break speed-reading records, we're trying to produce clear, readable documentation.

If your department produces long, detailed documents, you'll need to go with a longer line length, simply to fit your content on a reasonable number of pages. If most of your documents are short, and you believe your end user would benefit from a shorter line length or a larger typeface, choose a page layout with a shorter line length. The bigger the font, the longer your line can be.

Just My Type: Mastering Typography

Choosing the right typeface is one of the most important decisions you will make. With such a huge range of interesting typefaces to choose from, what's a tech writer to do?

In the world of typefaces, like that of many worlds, moderation is key. "Interesting" should not be among the characteristics used when choosing a typeface for technical documentation.

Typefaces used in your documentation should be easy to read and should be available to your end users. What do I mean by available? You should be confident that the fonts you are using are installed on their computers. If you specify an unusual typeface on a website or a Word document, it will look great on your own computer, but on your user's computer, it might display as something a bit more common, like Arial or Helvetica or Times. Even some of the typefaces that are packaged with a current operating system are not always available to users who have older operating systems.

Comic Sans is a typeface modeled on lettering used in comic books. Its broad appeal has caused it to be used as a typeface in some odd places, giving rise to the humorous anti–Comic Sans movement.

Websites such as **bancomicsans.com** ("Putting the sans in Comic Sans") and **comicsanscriminal.com** arose to prevent misguided "designers" from using it incorrectly. Not the best choice for *your* documentation!

For that reason, I like to use typefaces that I know all my users have and that look good on both PCs and Macs.

Use a limited number of typefaces and make sure they are clear and easy to read. Now is not the time to try out Comic Sans or Freestyle Script. You really need only three different typefaces for your technical documents:

- Body text
- Heading text
- Command line samples, code, and system messages

Body Text

Body text makes up the bulk of the documentation, and should be big enough and clear enough to read easily. Even though it's possible to enlarge PDFs or browser views, you don't want to force your user to have to do that.

Type size is normally measured from the top of the *ascenders* (the highest part of a character, the part that sticks up on a lowercase "h" or "k") to the bottom of the *descenders* (the bottom of a lowercase "y" or "p"), although today's digital type does not adhere to that measurement as strictly as old-fashioned typographers had to. (Sometimes digital type designers build extra forced space into their typefaces and consider it part of the size.)

Serifs are the short cross-lines at the ends of the main strokes of a letter in a typeface like Palatino, Georgia, or Times New Roman. A typeface without serifs is called "sans serif." (*Sans* means *without* in French.) Common sans serif faces are Arial, Helvetica, and Gill Sans.

Print type is measured in *points*. A single point is 1/72 of an inch. So when you hear of "12-point type," it means that there is a distance of 12 points between the top of the "h" to the bottom of the "y."

The *x-height* is the height of a basic lowercase letter that has no curves, ascenders, or descenders—typically, the "x," although a fancy typeface might do something different with its "x." The x-height of one font can be much smaller than that of another and give the appearance of being smaller overall, even though they are equally high from the top of the ascender to the bottom or the descender.

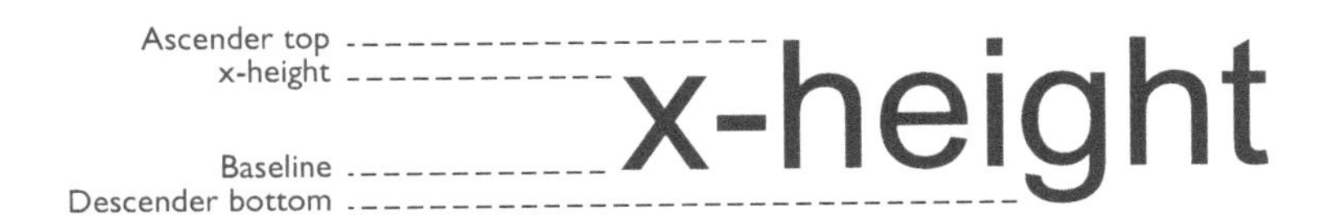

Ascenders and descenders determine the top-to-bottom size of a font. X-height affects its look and how much space it takes up on the page.

Type in print

Body text in print documentation should be no smaller than 10 points and no larger than 12. Most typefaces look good at 11 points. Some with a large x-height look fine at 10 points.

It is widely believed that serif faces such Times, Palatino, and Georgia are preferable for print documentation and sans serif fonts such as Helvetica, Arial, and Verdana are preferable for online documentation. This is not necessarily true—outside the US, books are commonly printed in Helvetica, and screen resolutions are so good nowadays that serif typefaces are much easier to read online than they were in the past—but to be safe, let's continue to follow those rules and select a serif face for print body text and a sans serif for browser text.

When choosing a typeface for technical documentation, think about the areas where a typeface can cause confusion: in the differences between a one and a lowercase L, and between a zero and a capital O. If your documentation includes a lot of numbers, you might want to make sure to select a face that has clearly readable ones and zeros.

Palatino: This is a 1 (one). This is a lowercase l (L). This is a 0 (zero). This is a capital O. This book is set in Palatino.

Georgia: This is a 1 (one). This is a lowercase l (L). This is a 0 (zero). This is a capital O.

Times New Roman: This is a 1 (one). This is a lowercase l (L). This is a 0 (zero).This is a capital O.

Arial: This is a 1 (one). This is a lowercase l (L). This is a 0 (zero).This is a capital O.

Verdana: This is a 1 (one). This is a lowercase l (L). This is a 0 (zero).This is a capital O.

Gill Sans: This is a 1 (one). This is a lowercase l (L). This is a 0 (zero).This is a capital O. This book's headings are set in Gill Sans.

Some serif and sans serif typefaces

Professional typesetters differentiate between *typeface* and *font*. A typeface is a single set of characters. Times Roman and Arial are typefaces.

Font refers to a single size and style of a particular typeface; therefore, while Arial is a typeface, Arial bold 10 point is a font. In today's world of digital typography, "font" is commonly used to refer to the typeface itself. Go ahead and use the common meaning unless you want to impress someone with your esoteric knowledge—or you're talking with a typesetter.

For body text in print manuals, I recommend you choose Palatino, Georgia, or Times New Roman. For headings in print manuals, I suggest Arial, although there is no harm in choosing any standard sans serif type, as long as it is easy to read.

On-screen type

Web fonts are measured in ems, percentages, points, and pixels and defined in the style sheet. If you do web design, you'll need to learn all about these units of measure. In the meantime, as a technical writer, the thing to remember is to make sure your font is big enough for anyone to read and that it's possible for the user to adjust the size if necessary. I recommend you use ems or percentages as your unit of measurement for type, which allow a reader to change the text size. Fonts look different on different computers and monitors of a different resolution, so it's important to let the user make the necessary adjustments.

For body text in web documentation, I recommend you choose Arial (substitute Helvetica if the user does not have Arial) or Verdana. (Verdana is an excellent face for online text, but it is very wide, taking up a lot of space compared to Arial and Helvetica.) Make sure your style sheet uses a non-fixed font of .875 em or 100%. Using percentages sets the body size at the user's default size.

Set your paragraphs to be *flush left, ragged right* (aligned straight only on the left side) rather than *justified* (aligned on both the right and left sides) and make sure you turn off hyphenation. Automatic hyphenation plays havoc with computer commands, file names, URLs, and product names.

Leading

Leading (pronounced "ledding") refers to the vertical spacing from one baseline of type to the next. You'll often see it called *line spacing* or *vertical space* in your publishing tool. Single spacing (for example, 12-point type on 12 points of leading) means that there is no extra space between the descender of one line of type and the ascender of the following line. Single-spacing is tight and hard to read in many typefaces.

Lorem ipsum has been used as dummy text in the printing industry for more than 400 years. It has roots in a piece of classical Latin literature from 45 BC.

This text is used when a designer wants to show what the visual elements look like without using real content. You can create your own Lorem Ipsum text from one of the many online generators such as **blindtextgenerator.com/lorem-ipsum**

The following paragraph shows you what this book's body text looks like in a single-spaced paragraph, using *Lorem ipsum* dummy text. While single spacing is fine for short passages, it can cause reading difficulty when used excessively.

> Lorem ipsum dolor sit amet, consectetuer adipiscing elit. Aenean commodo ligula eget dolor. Aenean massa. Cum sociis natoque penatibus et magnis dis parturient montes, nascetur ridiculus mus. Donec quam felis, ultricies nec, pellentesque eu, pretium quis, sem. Nulla consequat massa quis enim. Donec pede justo, fringilla vel, aliquet nec, vulputate eget, arcu. In enim justo, rhoncus ut, imperdiet a, venenatis vitae, justo. Nullam dictum felis eu pede mollis pretium. Integer tincidunt. Cras dapibus. Vivamus elementum semper nisi. Aenean vulputate eleifend tellus.

I recommend that the leading in body text be between two and four larger than your type size, depending on the typeface used and the amount of text on the page. Fonts with low x-heights, as discussed earlier, often look better with smaller leading; fonts with large x-heights need a much bigger leading. If you aren't sure what looks best, add three extra points; it will be fine.

Three extra points of leading means that 11-point type is on 14 points of line spacing, or leading. Every line in a paragraph will be 14 points below the preceding one. If you are manually marking up copy, you write it as 11/14.

The larger the font and shorter the passage, the less vertical line space you need—for example, a 36-point heading works fine single-spaced. Sometimes with large fonts, you can even choose negative line space, such as 36/30, a trick that is frequently used in graphic design.

In addition to standard line space, your desktop publishing tool or CSS styles allow you to add still more space before or after any given component. Suppose you choose to add 6 points after every paragraph. That means that as soon as you press the Enter key to create a new paragraph, your 11/14 paragraph adds six points more of blank space before starting the next paragraph.

I suggest you add six points after every component in a document, and zero points before all paragraphs and list items in body text. This gives you consistency throughout the documentation. You can add more points *before* major headings—enough so that the heading is farther away from the preceding paragraph than it is to the paragraph that follows.

Headings

Headings should be done in a typeface that is distinctive from the body, making it easy for the reader to scan the document and get the gist of it from the headings alone. A bold sans serif font is frequently used for headings.

Decide how many levels of headings you want in your documents. Five is typical, with the top level being the chapter title. Make sure the heading levels are sized differently and perhaps indented differently (as long as they follow the grid!) so they can be clearly distinguished from each other.

Chapter Title 28 points

Heading Level 1, 18 points

Heading Level 2, 14 points

Sample heading sizes in Arial

Command Lines, Code Samples, and System Messages

If you are responsible for any kind of developer documentation, you should use a special font for system messages or command line input. Technical writers usually use a *monospaced* typeface (meaning that each character takes the same amount of space as another) for this. In a monospaced font, a lower-case "i" or a blank space are the same width as an upper-case "W" or "M." This is opposed to a proportional font, in which each letter uses a space appropriate to its width. Monospaced code samples and commands mimic the way they look when a programmer types them in a command-line environment, making it easier for you to copy/paste real code and a user to correctly replicate your examples.

Consolas: 11 point Consolas. This is a 1 (one). This is a lowercase l (L). This is a 0 (zero). This is a capital O.

Courier New: 11 point Courier New. This is a 1 (one). This is a lowercase l (L). This is a 0 (zero). This is a capital O.

Monospaced typefaces

Monospaced fonts include the Consolas and Courier families. Convention is to use a regular font for system messages and **bold** to indicate user input.

Color Me Interesting

One of the great advantages of web-based and PDF documentation is that you can use as much color as you want.

Since your screen shots are almost certain to be full color, make sure any line illustrations you provide in the same documentation are also color. You can also enhance the text with a second color (that is, other than black). If you're not sure what color to use as your second color, refer to your company's branding guidelines and choose one of the approved colors.

Use color sparingly. Just because you *can* use color for every heading, bullet, number, and rule doesn't mean you *should*. Sometimes just using color for accents is enough—such as bullets, rules, the numbers in the chapter titles and numbered lists, and icons for cautions and warnings.

Although color is nice as a design element in illustrations, documentation, or websites, don't use it as an indicator, or the only indicator, for anything important. (I sometimes see illustrations in documents with huge color-coded charts, each color signifying something within the illustration.) Many people can't see the difference between colors, especially red versus green, so instead of depending upon color for meaning, use cues such as underlines or numbered callouts to supplement color. A good guide for choosing colors is how they print in black-and-white. If you can't see the difference between colors when they are printed in black-and-white, assume that others won't see the difference even when they are printed in color.

Ask at work if anyone has color deficiency. When you find someone, use him (because this is a gender-linked condition, it is likely to be a *him*) as a test subject. Although there are many degrees of color deficiency and your test subject may not cover every possibility, it's still a good way to help ensure that you have not assigned a significance to color that can't be understood by all.

The Final Layout

After you have chosen your fonts and made decisions about the margins, it's time to flesh out the page layout. The example page layouts on pages 251 and 252 follow the recommendations in this chapter:

- All pages in the layout align to the grid on page 242. The page is designed to fit on both US Letter and A4 sizes, although its primary use is US Letter.
- The body text is 11-point Palatino on 14 points of leading, flush left and unhyphenated. The line length is less than 75 characters. Headings are Arial.
- There are six extra points after each paragraph and list item. The headings have different amounts of extra line space preceding them, depending on the heading level.
- Arial 10 point type is used for all other text elements: figure and table titles, table content, footers, headers, and callouts.
- The index, not shown in the illustrations, would also use 10-point Arial on a two-column page. Don't forget that an index (as well as the table of contents) must also follow the grid.

The sample pages work if you want to use the same layout for both left and right pages, since it is centered on the page.

You can take ideas for your own template from the examples shown here, or modify one of the templates that come with your desktop publishing software using the guidelines in this chapter.

We've barely skimmed the possibilities of document design, but at least you've got some of the big-picture concepts and vocabulary to help you on your way. If page design and template creation interest you, you'll be glad to know that template design and web page design are part of many tech writing jobs.

If page layout is not your cup of tea, just remember that less is more. Limit your use of typefaces, sizes, and colors, and look for consistency.

Be consistent in your use of measurement increments—pick a width of a quarter-inch, for example, and repeat that width everywhere—between columns, as the second-level indent, or as a width for icons. (You can see that in the grid on

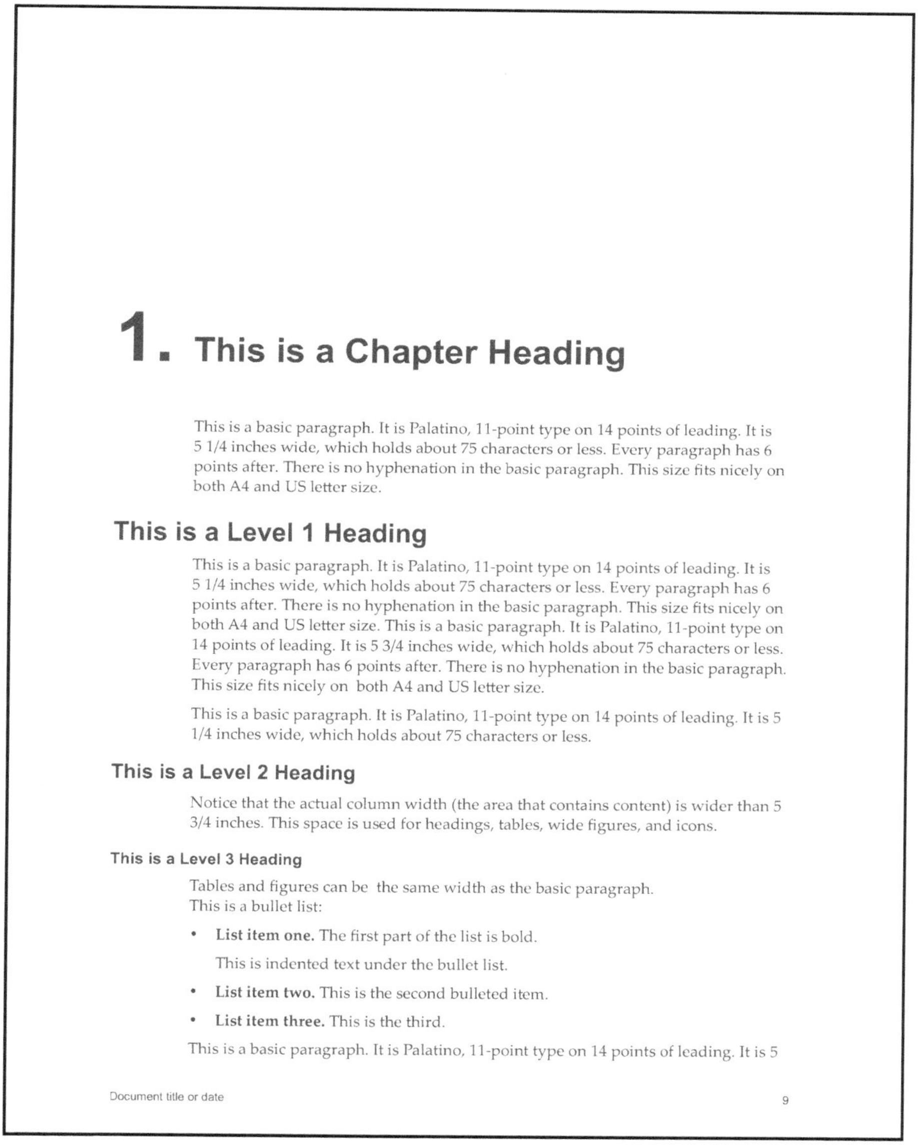

1. This is a Chapter Heading

This is a basic paragraph. It is Palatino, 11-point type on 14 points of leading. It is 5 1/4 inches wide, which holds about 75 characters or less. Every paragraph has 6 points after. There is no hyphenation in the basic paragraph. This size fits nicely on both A4 and US letter size.

This is a Level 1 Heading

This is a basic paragraph. It is Palatino, 11-point type on 14 points of leading. It is 5 1/4 inches wide, which holds about 75 characters or less. Every paragraph has 6 points after. There is no hyphenation in the basic paragraph. This size fits nicely on both A4 and US letter size. This is a basic paragraph. It is Palatino, 11-point type on 14 points of leading. It is 5 3/4 inches wide, which holds about 75 characters or less. Every paragraph has 6 points after. There is no hyphenation in the basic paragraph. This size fits nicely on both A4 and US letter size.

This is a basic paragraph. It is Palatino, 11-point type on 14 points of leading. It is 5 1/4 inches wide, which holds about 75 characters or less.

This is a Level 2 Heading

Notice that the actual column width (the area that contains content) is wider than 5 3/4 inches. This space is used for headings, tables, wide figures, and icons.

This is a Level 3 Heading

Tables and figures can be the same width as the basic paragraph.
This is a bullet list:

- **List item one.** The first part of the list is bold.

  This is indented text under the bullet list.
- **List item two.** This is the second bulleted item.
- **List item three.** This is the third.

This is a basic paragraph. It is Palatino, 11-point type on 14 points of leading. It is 5

Document title or date 9

This chapter title page layout adheres to the grid.

page 242.) For vertical measurements, use a measurement of six points to append to a paragraph tag and use that all the time.

If you stick with the grid and follow the guidelines in this chapter, you're bound to have a document that looks professional and is readable. A good solid design can turn "blah" documentation into really good documentation.

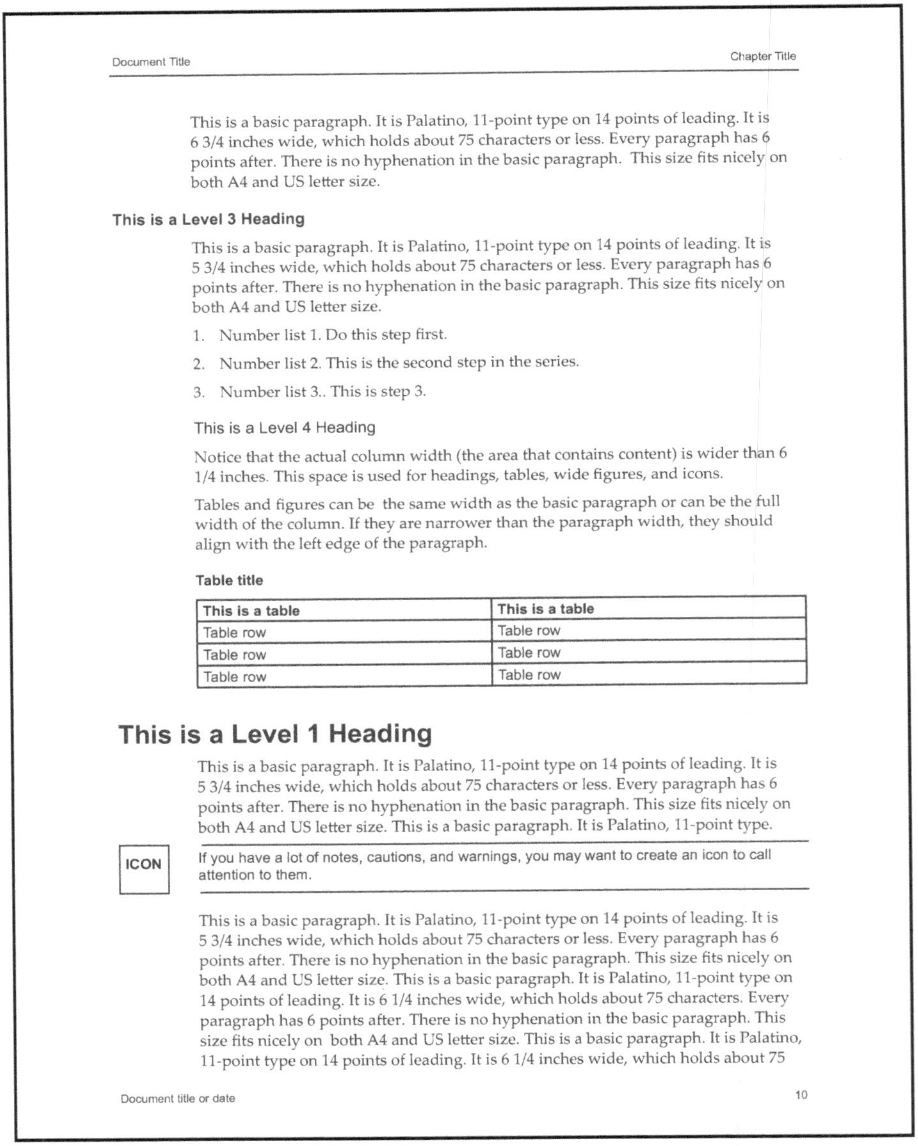

Document Title | Chapter Title

This is a basic paragraph. It is Palatino, 11-point type on 14 points of leading. It is 6 3/4 inches wide, which holds about 75 characters or less. Every paragraph has 6 points after. There is no hyphenation in the basic paragraph. This size fits nicely on both A4 and US letter size.

This is a Level 3 Heading

This is a basic paragraph. It is Palatino, 11-point type on 14 points of leading. It is 5 3/4 inches wide, which holds about 75 characters or less. Every paragraph has 6 points after. There is no hyphenation in the basic paragraph. This size fits nicely on both A4 and US letter size.

1. Number list 1. Do this step first.
2. Number list 2. This is the second step in the series.
3. Number list 3.. This is step 3.

This is a Level 4 Heading

Notice that the actual column width (the area that contains content) is wider than 6 1/4 inches. This space is used for headings, tables, wide figures, and icons.

Tables and figures can be the same width as the basic paragraph or can be the full width of the column. If they are narrower than the paragraph width, they should align with the left edge of the paragraph.

Table title

This is a table	This is a table
Table row	Table row
Table row	Table row
Table row	Table row

This is a Level 1 Heading

This is a basic paragraph. It is Palatino, 11-point type on 14 points of leading. It is 5 3/4 inches wide, which holds about 75 characters or less. Every paragraph has 6 points after. There is no hyphenation in the basic paragraph. This size fits nicely on both A4 and US letter size. This is a basic paragraph. It is Palatino, 11-point type.

ICON — If you have a lot of notes, cautions, and warnings, you may want to create an icon to call attention to them.

This is a basic paragraph. It is Palatino, 11-point type on 14 points of leading. It is 5 3/4 inches wide, which holds about 75 characters or less. Every paragraph has 6 points after. There is no hyphenation in the basic paragraph. This size fits nicely on both A4 and US letter size. This is a basic paragraph. It is Palatino, 11-point type on 14 points of leading. It is 6 1/4 inches wide, which holds about 75 characters. Every paragraph has 6 points after. There is no hyphenation in the basic paragraph. This size fits nicely on both A4 and US letter size. This is a basic paragraph. It is Palatino, 11-point type on 14 points of leading. It is 6 1/4 inches wide, which holds about 75

Document title or date | 10

Interior page, based on the same grid.

Chapter 21

Gaining a Global Perspective: Localization and Translation

Going international is important for your company. Are you equipped to handle it?

What's in this chapter

- Finding a translation company that can keep up
- The difference between localization and translation
- Improving base documentation to make it easier to translate
- Dealing with the scheduling challenges that accompany translation

---

As your company goes global, you're going to need to think about the fact that at least some of your documentation will have to be translated into other languages. In some organizations, the Tech Pubs department is responsible for managing all translation for the company.

Translation can be very expensive. You can help keep costs down by applying best writing practices and tight scheduling.

This chapter helps give you the tools you need to write documentation that can be easily translated, and to work successfully with translation companies.

Finding a Good Translation Company

The best thing you can do to ensure that translation goes well is to find a good translation company. Before you start a movement to hire a translation company, find out if any department in your company is already using one. It is not uncommon to find out that many different people in the company are already using translation services.

If your company does not have a translation firm it works with, then it's time to get one. You might bring together all the departments in your organization that use or might use translation services and go through a bidding process to find a vendor.

One Can Be Better Than Two

It's best to have one translation vendor at your company. For one thing, translation companies will often give a discount to companies that guarantee them a minimum amount of business. Even more important, using the same company means that language and terminology will have a consistency throughout all of the translations. In keeping with my often-repeated theme of the importance of consistency, it's important for an item to be called by the same name in every document.

You might be surprised to discover that very few translation companies keep staffs of translators on site. They use freelance translators from around the world, and these freelancers often work for many different companies. A lot of the value in an individual company comes from the relationship that is built between your organization and the people at the translation company.

When the Marketing department is using one translation company and Tech Pubs is using another and Business Development is using still another, and Sales is asking the people in the local offices to translate material themselves, it's bound to be bad in the end for the customers as they struggle to understand whether a term in a data sheet means the same thing as a term in the user guide.

The frequent exception to the "one translation company" rule is for legal translation. Corporate legal services usually use translation firms that specialize in legal translation. The translation is done or reviewed by lawyers who understand the law in the countries for which the material is translated.

The rest of us have different needs, and should look for a translation company that can meet them.

The Questions to Ask

When considering translation companies, ask the following questions:

- Can the translation company handle not only the amount of work you have now, but also the work you'll have in next few years? If your international business is expanding, you'll need a company that can keep up with your growth.
- Can the company work with the schedules you'll give them? If you have frequent needs for very fast turnaround, you might want to work with a small company that doesn't have a lot of project management.
- Does the company have access to skilled translators in the languages you most use?
- Will the company let you meet the account representative or project manager who will be handling your business? Make sure this person is someone you like and can comfortably communicate with.

> **INSIDERS KNOW**
>
> When choosing a translation vendor, cost matters, of course, although the price per word does not always differ as hugely as you might expect from one company to another. Where you often see a difference is with project management costs.
>
> If your company does a lot of translation business and can guarantee a minimum amount of work, find a vendor who will give you a discount.

The Translation Process

As you talk to the representatives of the companies that are competing for your business, ask each to describe the process they use. It will be something like the following:

1. **Analyze the projec**t. The translation company gathers requirements and analyzes the material you provide.
2. **Plan the project.** The company prepares the files for translation and gets the right translators in place.
3. **Complete the translation**. This will cause the translation memory, a database of text segments that have been already translated, to be created.
4. **Perform quality checks.** A different translator reviews the job for accuracy.

Translation Versus Localization: What's the Difference?

You'll hear the terms "translation" and "localization" intermingled. *Translation* refers to the act of converting content in one language directly into content in another language.

Different countries have different ways of expressing large numbers. Forty-thousand is written as 40,000 in the US, 40.000 in many European countries, and 40 000 in others. Translators will handle this for you, but it's a good idea for you to know about numbers so you avoid any situations that can cause difficulty (such as including a number in a bitmap).

Localization, however, is the act of changing content so that it is relevant to the locale. That can mean changing details like measurements, money, time zones, and names. You might even find yourself changing the colors in your document design if it's determined that certain colors are not well-received in a given locale. Localized documentation should appear to have been developed in the country in which it is distributed.

Localizing content is not done automatically as part of the translation process. Localization is usually done in-house, but if you feel it is something you cannot do, discuss this need with your translation company.

Developing a Global Awareness

Ideally, technical writers should create documentation that doesn't need a lot of localization. This requires a global awareness that not all of us have at first, but as you acquire it, it reaps benefits for your company. You save much of the expense of localization, you avoid having multiple source files with different content, and you still make customers happy.

How do you achieve this? By writing source documentation that makes sense no matter where it is published.

When writing for an international audience, it's more important than ever that you understand the product you are writing about. If you don't understand something, you can't describe it clearly, and this shows up during translation.

Some authoring environments provide support for localizing numbers and dates by letting you use a standard markup that will be converted to the appropriate format for whichever locale you choose. Ask your tools support people about localization.

If you don't have that kind of support, or if you are asked for advice, here are some good rules to help avoid localization problems:

INSIDERS KNOW

Translation memory is a database of text segments that have been already translated. The source content and its corresponding translated language are stored together. The translation memory is used with each subsequent job and applied to segments that already exist in the memory. The more memory that builds up, the less expensive future jobs become, as more and more of the job is "pre-translated."

Trados is a common translation memory tool used by many companies. Ask yours to show you how it works. Understanding translation memory will give you a better sense of how to prepare your content.

- **Make sure that you publish local telephone numbers as well as toll-free numbers.** Many North Americans don't know that toll-free numbers do not work around the world. If you put your company phone numbers in the documentation, include both local and toll-free.
- **Avoid confusing date formats.** Dates differ from one locale to another, and users won't know if 2/8/13 means August 2, 1913, or February 8, 2013. One way to avoid this is to spell out the names of months and use all four digits for years. I like the format dd-Month-yyyy, or 08 February 2013.
- **Avoid mentioning money and place names.** It's a lot easier not to talk about those things at all than to go through each document and change the references.
- **Avoid locale-specific names.** If your screen shots contain sample input with user names and addresses, use a mix of locale names or no place names if you can. If you must include names and addresses, avoid humorous references like Star Trek or Disney character names. Mickey Mouse is not called "Mickey Mouse" throughout the world!
- **Avoid locale-specific measurements.** When giving measurements, provide both *imperial* (that is, the system used in the US with inches and feet) and metric. Don't use abbreviations for measurements, either, such as " or *in* for inches and ' or *ft* for feet.

The Translation Glossary

Besides writing source documentation that avoids locale-specific references, you should work with your translation company to create a glossary. A translation glossary is different from a standard glossary. Instead of defining the mean-

As you work more with translation, you might see and be puzzled by the industry abbreviations L10N and T9N. *Localization* is shortened to L10N to indicate that it starts with "L" has 10 letters in between, and ends with "N." *Translation* starts with "T" has 9 letters in between, and ends with "N" to result in T9N.

ings of terms, it defines the translations of terms into each language in use.

The glossary, when used consistently, means that every time a term appears in the original language, it is always translated to the same term in the target language. In addition, the glossary states which terms are *not* to be translated; for example, brand names are typically not translated.

Accepted abbreviations need to be listed in the glossary as well—remember that an acronym in English will not be the same in a different language, where it stands for a different set of words.

It is important to keep the glossary maintained. With each new project, the translators working on your jobs should enter new terminology. As the client, it's your responsibility to regularly review the glossary with the international teams in your company to make sure that the terminology used in each language makes sense.

By the way, this is where an XML schema like DITA really shows its value. Having reusable content chunks that use the same wording is great for translation. Once the content gets into translation memory (TM), it can be reused again and again without re-translation. However, every time a paragraph is copied, tweaked, or modified in any way, even if it was already translated, it has to be looked at again and translated again before it is added into the translation memory.

Writing for Translation

In addition to making your content locale-neutral, there are other things you can do to best prepare documentation for translation. Those things might not be what you think. You might have heard that you should avoid slang, contractions, and passive voice when preparing for translation. The truth is you should probably avoid those things even if your documentation is not being translated. If you adhere to good technical writing practices as described in this book, your content should be ready for translation.

There are three areas you really need to concern yourself with when preparing documentation for translation:

- Make sure there is enough space for language expansion.

- Pay attention to formatting that can affect the translators' output.
- Keep text separate from bitmaps.

Language Expansion

Different languages take up different amounts of room, and it's not always easy to remember that when planning for translation. If the source file is a PDF manual or HTML content, it probably doesn't matter if the target language extends the content by 30, or even 50 percent. But if there is limited space, you must keep a very close watch on the translated content and you might have to work with the translator to shorten it.

Some examples of where this happens are unexpandable desktop software user interfaces or labels on image-based buttons or menus. I remember a situation where a button on a website was designed to precisely fit the size of the English word "Go." As you can imagine, that caused problems, since the word is much longer in every other language..

Allow an extra 30 percent on average for translation, although some languages can be longer or shorter.

Watch Out for Formatting

Some of the trouble areas in source documentation occur when writers don't properly use the paragraph tags and styles that should be set up in the template. Keep an eye out for these: formatting issues

- Use the styles and tags defined in the templates and avoid manually applying formats and fonts.
- Make sure all commands, path names, user input, and code samples are properly tagged. If you leave some commands or code in your regular body font, a translator is likely to translate some of it, not realizing its significance.

One unexpected advantage of using a different typeface (such as Consolas or Courier) for system messages and user input is that a translator knows not to translate anything that's in this special font. You don't want the translator to accidentally change computer commands.

- Avoid using spaces, tabs, or hard line breaks to create tabular formats. Instead, use tables.

- Don't use a fixed row height in any table or a fixed-sized text box anywhere. If the translated content exceeds the original content, it will be hidden and you might not know it.
- Make sure you do not change tables of contents or indexes or modify a PDF after they have been generated. The translator won't know about your changes.

Keep Your Text Editable in Graphics Files

Remember the image file types discussed in Chapter 12, "You Want it *How?*" Raster files (bitmaps) create their own set of challenges when working with translation.

Image files

Make sure that your image (raster) files don't include text. You can tell if this will be a problem if the file extension is .jpg, .tif, .bmp, or .gif, and there is any kind of text within the picture that requires translation. Unless you have the original layered file in its native Photoshop or similar form, the graphic file will have to be recreated by the translator. It's very expensive for a translator to redo a graphic of this kind.

If you have raster images that include text that *must* be translated, do your best to provide the translator with the original version, done in Photoshop or a comparable program. (The translator may ask for *layered* files; that means that the original file contains the image on one layer and the type on another. The image is flattened for final production. The translator wants to work directly on the text layer.

Avoid problems by always using images that are text-free. In the documentation, put callouts in text boxes within the figure's frame.

Screen shots

If the UI of the software itself is being translated, you will have another set of challenges when creating screen shots. There is going to be a definite time crunch, once you reshoot the newly translated screens, to fit them into documentation.

This is where having size standards for graphics helps a lot. If all of your screen shots are the same size and same resolution, the newly translated ones can automatically replace the shots in the existing documentation. Just make sure that the new screen shots have the same file names as the previous screen shots.

Line drawings

Line illustrations (vector files) are much easier for a translator to manage, because they can edit directly in the file. Provide the vendor with original files with .ai, .eps, or .svg file extensions.

The .svg format is recommended if you are using XML for content. The text in an .svg file will be extracted and presented to the translator as plain text. Once translated, the text will be inserted back into the graphic.

Managing the Schedule

It's not just the translation company that will be doing all the work. On your side, you've got to produce documents that can be translated. That's not so hard—translators can handle almost anything you throw at them. If you follow the guidelines in this chapter, the source material will be in good shape and ultimately you will save money on translation.

The harder job for you will be managing the schedule to make sure that translations are finished when you need them. This means running a very tight ship and pre-planning every step of the way while working closely with the translation company. The scheduling tips discussed in Chapter 11, "You Want it *When?*" will come in handy, but even they might not be enough as you deal with the special challenges that come with managing translations.

Work with the Vendor

Until you gain a good understanding of how long a translation job might take, you need to work with your translation vendor each time you give them a project. If you're not in a big hurry, a translation company will not do a rush job. If you're in an enormous hurry, a good company can pull out all the stops to deliver (and be prepared to pay for the service).

Translation planning should be part of the document plan discussed in Chapter 9, "Process and Planning." Work with your project team to determine when you need to deliver translated documentation and then plan accordingly.

In some businesses, translated material can wait until after the initial release. In a situation like that, the company finishes the documentation and distributes it to customers, and then starts on the translation leg of the journey. Four or five weeks later, it releases localized software and documentation in the other languages needed for business.

However, you might work for a company that needs its translated documentation at the same time as its local-language documentation. For example, if you work for a consumer electronics company that includes a printed user guide in the box, that guide is likely to be in several languages, all of which have to be printed at the same time. Or that same consumer device might have help provided in several languages and built into the firmware. Even if your business allows the additional time after an initial release to have a second release of translated material, you still need to finalize the documentation to match the newly translated software.

Extreme Scheduling

When you are required to produce translated content at the same time as local-language content, that's when you will really have a chance to flex your project management chops.

I contact the translation company as soon as I can and send them as much information as I have at that time: target languages, drop-dead due date, and estimated word count. I usually send draft versions of the items to be translated and tell them to plan for, and give me an estimate for, an expansion of 10 to 20 percent. This allows me to open up a purchase order.

The local-language content must be completed well before the product is released to allow time for translation. I work with the translation company to determine the last date by which I can give them frozen source files and still have finished translation before the release date. There is little allowance for change.

As soon as the draft is solid, I send the files to the translation company. Although I will likely have to provide additional content and changes closer to the due date, and will pay for those changes, this lets the translators do the bulk of the work and enter terms into the translation memory. With luck, the last round of work becomes a refinement stage instead of a major translation effort at the last minute.

The More Ways You Have to Say Thank You, The Better

Whether you say *thank you*, *gracias*, *kiitos*, or *merci*, knowledge of translation management is something to be thankful for. It's a good skill to add to your bag of tricks, even though it certainly has its own challenges. Best of all, it helps you be involved in the growing customer base of your company. The more languages being translated, the more worldwide business your company is doing, and that's always a good thing.

Part 6. I Love My Job, I Love My Job, I Love My Job...

In this section, we'll look at tech writing from some unexpected angles. There will be days when you celebrate the fact that you are lucky enough to have such a great job, but there will be other days when you have to remind yourself of why you come back to work each day. Tech writing has its own unique set of challenges.

I think you'll find that you'll be happy more often than not. To help you make the most of your career, *The Insider's Guide* wraps up with insights and advice about planning where you want your tech writing career to go.

Chapter 22

Working Outside the Box

Consulting, contracting, telecommuting, and the issues that go with working outside the office.

What's in this chapter

- Work-from-home benefits and headaches
- Staying in the loop when you're no longer down the hall
- Consultant versus captive—which one looks good to you?

Not only do people no longer expect or even want to stay with a single employer until retirement, now they don't even assume they'll come to the office every day. More and more people are exploring different employment models.

In this chapter, we'll take a look at work options you might be considering—or might not have thought of before. Kick off your shoes and get comfortable.

Working From Home

Working from home has become so common, it's acquired its own acronym: WFH.

Of course, with today's technology, it's easy to work off site. Many people communicate by email and instant messaging throughout the workday, even with their next-door cube neighbors. It's not a huge transition to communicate in the same way from a remote location. If a worker has an emergency, has to go out of town, or is waiting for a repair person, employers are generally fine with the employee working from home during that period.

And then there are people whose employers requires them to regularly work from home offices. Some companies have eliminated permanent stations for many of their employees; others have projects with so many employees scattered across different locales that there's no true central office. Some workers live so far from the office that they have an arrangement with their management to permanently work remotely.

Can You Really Get Work Done While In Your Bathrobe?

Can a person be productive working from home? The short answer is yes. A lot of technical writers get a lot of work done while wearing fuzzy slippers. Writing is a solitary task and requires great concentration. When the information-gathering phase has wound down, many writers feel they can be much more productive without the distractions of the office.

Before you go shopping for loungewear, however, there are some things about working from home that you should understand.

First, it takes discipline. Because there's no one to "crack the whip" over you, you have to do it for yourself.

That doesn't mean you can't throw in a load of laundry while you're working. But it does mean you have to resist the temptations that beckon. It's often much easier to be distracted by what's going on at home—children, pets, a back yard swimming pool, housework that needs to be done—than what's going on at the office. And the home worker is often an easy mark for neighbors and family members who assume that because she is at home, she's available for visits or to do favors.

The number of people who telecommute increases every year. The Telework Research Network estimates that about 20 to 30 million people currently work from home at least one day a week. See **teleworkresearchnetwork.com** for more statistics.

It's a good idea to create a clear boundary between work and home life. Make sure your home office is its own room, or at least a section of a room with a divider to keep it separate. Walk into your office every morning, and plan to work. Come out during break and lunch periods and again at the end of the day. You'll get much more done than if you sit in front of the television with your laptop all day long.

When the Walls Start Closing In

Sometimes a writer, by choice or not, works from home full time. This can sound like heaven to some, but WFH does have challenges that can affect even those who work at home a couple of days a week.

- **Working at home can be lonely.** Even the most introverted tech writer needs some human contact; people are social creatures. In addition, you miss the office gossip, the social gatherings, and a good deal of the important news.
- **Your co-workers in the office don't always remember to dial you in.** The only way you can be successful while working remotely is if you attend all necessary meetings. Often, the people running those meetings from the office don't remember to make the conference call or set up the web meeting. You can sit on hold, be disconnected, or miss an entire meeting if the meeting's organizer doesn't remember you.
- **You miss the hallway conversations.** Yes, meetings are where a lot of things happen, but I think much more happens when you overhear random conversations, wander over to a colleague's desk to ask a question, or continue a discussion after the meeting is over. You risk missing a lot of that even if you work on site, just because a tech writer often works on so many different projects, but it's made much worse when you're working off site.
- **You can find yourself working too many hours.** When you work at an office, there's usually a natural stopping point—you get hungry, or tired, or have to catch your train. When you work at home, often there are no cues to say the workday is over. Sometimes people continue on without a break and don't realize how much time has passed. If you are in a different time zone from the on-site workers, you might be receiving email and messages and attending meetings well into the night.

The home worker can be even more susceptible to the down sides of working at a computer. Make sure you have a good chair and an ergonomic setup. Do regular stretching exercises. Stand up every half hour or so to keep your circulation moving. And give your eyes a regular break by moving them away from the screen and around the room.

Websites such as **safecomputingtips.com** contain some exercises that will help any computer worker.

There's No Place Like Home

With the right balance in place, working from home can be great. Walking down the hall is the shortest commute you'll ever have. There's nothing like looking out the window at snow or rain and knowing you don't have to go out in it.

And it should be fine for the employer, you might think. Arguably, you can get a lot more done if you're not spending a good portion of your time fighting traffic or fielding constant office interruptions. And your department and division are likely to have members working globally. Since managers deal with employees remotely all the time, it shouldn't be any big deal to have a few writers working from home.

Maybe.

WFH: The Other Side

Technical writers, more than others, often feel that an expectation of the job is to be able to work from home, if not daily, then one or two days a week. It is an expectation that many technical writers ask about during a job interview. Although a lot of technical writers might think that WFH is a job entitlement, the manager and the company don't always feel the same way.

Way back in Chapter 3, "Having the Write Stuff," I mentioned the importance of being part of the business. This means not only understanding what's important to your company, but also *acting* as if you're part of the company. If you expect to share in salary increases and bonuses, you have to give yourself some visibility. Look around you and get a sense of what the culture is, and if no one else in other departments is working from home, don't assume that being a technical writer makes you an exception. This is especially true if you work in an Agile environment, where face-to-face discussions and daily meetings are part of the work life.

Webcam meetings are a great way to stay in touch when you work remotely. Just remember your webcam manners and make sure you're dressed appropriately from the waist up and have combed your hair. Many remote workers keep the webcam on all day as they work, and tend to forget about it, much to their embarrassment when they're caught doing something they'd rather not have on camera.

In today's fast-paced high-tech world, a lot of work comes in unexpectedly and requires a fast turnaround. A manager often finds it a lot easier to discuss assignments face-to-face, and it can be frustrating to walk past a row of empty chairs looking for someone—anyone—to take on a project. Communicating with someone remotely means holding many email and instant messenger conversations, often repeating the conversations with multiple people, and that takes up valuable time.

It's also difficult for a manager to schedule meetings when people are in and out at different times. Looking at the calendar to find out when people are available, and if that doesn't work, setting up conference calls and online meetings, is a hassle. It might not seem like one to you, at the other end of the phone call or live meeting, but multiply this by many staff members and many projects, and the hassles start to build up.

If it's a hassle when offsite workers are trustworthy and put in their time, it's worse when offsite workers abuse the privilege. Many managers have been burned by offsite workers who don't do their fair share of work but make their slacking off very hard to prove. If you now report to a manager who has been through that, it can be difficult to gain that manager's trust if you want to work off site.

How to Work Off Site and Still Be a Team Member

If you work at home once in a while or often, it's important to maintain contact with the office to avoid the "out of sight, out of mind" syndrome. Even if your manager is fine with WFH, let's make sure to keep it that way.

Staying out of office politics can be one of the best things about being a contractor. It doesn't pay to get too hung up on the twists and turns of who's doing what to whom. Just keep track of who approves your time card and what that person wants you to do.

- **Communicate frequently and regularly.** That means always being available on instant messenger services and email during working hours. Not only available, but proactive in your communication. Copy your manager on any email that might be of interest. And don't be afraid to pick up the phone once in a while if there's something going on that might seem urgent.
- **Be productive.** Don't fall prey to the distractions of the home office. It's important to make sure that you produce at least as much as you do while in the office. Don't think of yourself as someone who works on single discrete projects—you need to participate in all the work that comes into the department, so it can take a bit more vigilance to make sure you're involved. If you have any down time or need a break from the project you're working on, ask

your boss if you can take on something else. Don't wait for him to have to ask. He may never ask if you're not there.

- **Make sure you're in the office when you need to be.** That means you should show up at work whenever your manager wants you there, even if you don't think it's necessary. If there are all-hands meetings or developer meetings or social gatherings, show your face often enough so that people think of you as a regular fixture.

  If you're a true offsite worker, meaning you have no permanent desk at the office, and you live close enough to the office, make sure to schedule meetings and lunch with people from the office upon occasion. That gets you dressed and out of the house and gives you more of a sense of still being part of the company.

If you are a remote worker who cannot easily get to the office, make sure you stay in touch regularly and that people know your face. Make liberal use of your webcam and conferencing software when you talk to co-workers. Fly or drive in as often as you can and stay long enough so people get to know you; then hold face-to-face webcam meetings and conversations as soon as you return to your home office so they feel as if you are still there.

If you work on site now and think you'd like to work from home some or all of the time, start out slowly and build the trust with your manager. Once you have proven yourself as a valued employee, you have a much better chance of being allowed to telecommute.

Start by working from home one or two days a week, in the middle of the week. Important things happen on Mondays and Fridays and being "gone" one of those days can create the perception that you're off for a long weekend, even though you know that you're busy working.

As you follow the guidelines in this chapter, it's important to remember that even though you're not sitting with the rest of your team members, you are still a member of the team. As you continue to act like one, before long, the day will come when people don't even realize you're not in the next cubicle.

If you decide to become a true free agent, make sure you consider all that is involved in self-employment. You might need to consult not only a lawyer, but an insurance agent if you are not covered through a family member. Also make sure you clearly understand how self-employment affects your tax status.

Consultant or Captive?

After you've built up your job experience and expertise, you might start to wonder whether it's really necessary to work for someone else at all. Many technical writers have a lucrative and enjoyable career as free, or somewhat free, agents. However, as true independent workers they may be on their own for health insurance, retirement savings, and setting aside the right money for all kinds of taxes.

You might refer to an independent worker as a consultant, contractor, or freelancer. None of these is a payrolled full-time employee of the companies for whom they actually do the work. In fact, they can work for more than one company at the same time.

You Say Consultant, I Say Contractor

The terms in this chapter mean different things to different people, and in your part of the country or world, they might be used differently as well. The important thing for you, as a worker, is to determine if any of these lifestyles appeals to you. If they do, you might want to find a way to make it happen.

A *consultant* is typically a person who comes into another company to solve a problem. The consultant might be a full-time employee of a consulting company, or might be self-employed. Often, consultants receive a set fee for a project, since they are selling a solution rather than their time.

A *freelancer* is another type of independent worker, usually hired to work on a single project on an hourly or per-project basis. As totally independent worker, freelancers work off site on their own equipment, coming in to the office for meetings when necessary.

A *contractor* fills a job temporarily, and bills for time worked. The contract defines the length of time and payment on an hourly, or less often, daily, basis. The contractor often functions like a full-time employees, working on site and on the same projects during the standard workday.

If you become self-employed, you must adjust your billing rate. You will become responsible for many more of your own expenses. You also are not guaranteed continual work throughout the year.

A rule of thumb is to aim for an hourly rate of 1/1000th of your yearly salary, so if your current salary is $60,000, you would charge $60 an hour. Another formula suggests you charge 1/100th of a yearly salary as a daily rate, which would be $600 a day.

The rate you can charge will also depend on the market and what businesses are willing to pay. It could be a good idea to discuss your plans with an accountant before going into an endeavor like this.

We discussed the hiring of contractors in Chapter 11, "You Want it *When?*" In this section, we'll look at it from the other side of the fence.

Free Doesn't Always Mean Independent

Not all contractors are true independent workers. When you work on a company's premises, use a company's equipment, and do the work the way the manager of the company tells you to do it, you are not independent.

Contractors in the technical writing world frequently work for agencies, or "job shops"—that is, companies that source temporary workers. The job shop manages the contract with the work site and pays the technical writer, who is actually an employee of the agency, deployed to the company that needs a technical writer.

Some contractors are able to gain more independence—and more money—by also owning a business that contracts directly with the company. If an opportunity comes up with a job shop, they take it. If an opportunity comes up to work directly, they jump on that.

In all cases, the bottom line is on a tax form. The government has its own ideas about what constitutes self-employment, and you must follow tax rules.

Worst—and Best—Of Both Worlds

You might be wondering why someone would want to work as a contractor. There's not a lot of security, you still have to follow the rules and regulations of the company for which you are performing the work, and you don't get such benefits as paid holidays and bonuses. You can have long periods of unemployment with no time or money to enjoy them, as you make the rounds looking for work.

Working as a contractor is a great way to get your foot in the door if you're looking for a permanent job. If that's your goal, make sure you let the hiring manager know you're interested if anything should come up. (It's not unheard of for a manager to hire a full-time employee without realizing that the on-site contractor was interested in the job.)

Be aware, however, that it is sometimes difficult for a company to hire a contractor because they have to pay a substantial fee to the agency the contractor works for.

Well, first of all, the contractor, as a temporary worker who steps in when times are tough, usually makes much more per hour than a full-time worker does. Contractors with specialized knowledge and skills often command very high rates of pay and are in great demand. And because a temporary worker is paid for all hours worked, the contractor can make quite a bit working overtime.

For many contractors, the enjoyable part of the job is that it offers variety. They like to move from job to job. They have a chance to try out many different companies and learn new skills.

And remember the discussion earlier in this book about the necessity of becoming involved with the business and what's important to it? A contractor doesn't have to do that. And for many, that's just fine. It means avoiding office politics.

Put Your Best Foot Forward

What makes a good contractor? The most important qualification is the ability to jump into a new situation without fear or self-consciousness. A technical writer should be able to "hit the ground running," as they say, without a misstep. Contractors are usually hired not only for their technical writing expertise and technical knowledge, but also because they know the tools that are in use in a given company.

If you like the company you're working for and it likes you, don't be surprised to receive an offer for a full-time job. As soon as a company gets funding to hire, the contractor is often the first one to receive an offer.

Don't limit your networking to organizations where you meet other technical writers. Look for any place where programmers, hiring managers, recruiters, and other professional people might gather. Any of them is likely to work at a company that hires technical writers.

Meetup, a website that helps groups of people with shared interests plan meetings and form offline clubs, often lists meetings for programmers and other people interested in technology. Go to **meetup.com** and sign up for any such groups that you think might fit in with your interests and expertise.

Set Me Free

If you don't want to be "captive" and would like to know how you might become a contractor, you'll be in good company. Many tech writers prefer being contractors. They like the idea of working at a variety of companies, they like the idea of having time off between contracts, and they mostly like the idea of commanding a higher hourly rate.

Chapter 11, "You Want it *When?*" talked about the thought process you might go through when hiring a contractor. Now put yourself on the other side of that transaction. Are you the kind of person a busy, short-staffed Technical Publications department would want? You might be a hard worker, a quick learner, and a joy to have on board, but if your resume doesn't reflect all those characteristics plus a skill set needed by a given company, it can be difficult to transition into contract work.

If you're a contractor or want to become one, make sure your tools and technical knowledge are extensive and up to date. Employers looking for temporary help want people who can jump in and use their software without any training time. When you're on a break between contracts, get up to speed on the latest and greatest help authoring tool, publishing tool, or markup language so you can add it to your resume.

Breaking into contracting uses many of the same methods discussed in Chapter 4, "Breaking Into the Field." Networking is key, and you must network continually, always on the lookout for the next job.

Many contractors get all their work by networking at professional organizations. The more technical writers you know, the more likely you are to be on their mind when they next have or hear about an opening. But it's not only technical writers that will help you. You need to think about how you can meet the other people who work in the companies that might hire you. You can meet these people any place—at your gym, in line at a fast-food restaurant, in the grocery store, or at the yearly science fiction convention. You need to be always ready to make a business connection.

Make sure you carry a supply of business cards with you at all times. And take a look at some of the tips in Chapter 24, "Managing Your Career." You'll find some other ideas for ways to promote yourself.

Chapter 23

I Didn't Think It Would Be Like This!

Unexpected quirks and hazards of tech writing and how to deal with them positively.

What's in this chapter

- The down side of the tech writing field
- Avoiding that overwhelmed feeling

Every job has some difficulties, and tech writing is no exception. As great a career as tech writing can be, I know only too well that the field has hurdles that you won't find in any other.

In corporate-speak, they aren't problems, they're "challenges," and sometimes it's true that how you look at them, think about them, and respond to them makes a big difference in how crazy they can make you. In this chapter, you'll get what I hope is an honest discussion of some of the challenges—and some of the solutions—that you might encounter in your tech writing career.

But don't worry if you find this chapter discouraging. In Chapter 24, "Managing Your Career," you'll learn ways to counteract some of the challenges by making the most of the opportunities available to you.

The "Dark" Side of Technical Writing

Professionals in today's high-tech industries face a few problems along with the pleasures. It's not all foosball and free espressos. The excitement and fast pace of the computer industry can also mean unrelenting pressure and the stress that comes with it. The tight deadlines and heavy demands of release dates often result in long hours of relentless work.

As if your tired, scratchy eyes and tension headaches aren't problem enough, many software programs with heavy mouse use can cause carpal tunnel syndrome or repetitive stress injuries. *Dilbert* cartoonist Scott Adams wasn't kidding when he said that technology is no place for wimps!

Choose Your Deadline: Aggressive or Insane?

Those don't sound like very appealing choices, but sometimes they're the only ones you get to pick from. It's a hazard of high tech, not of technical writers specifically—the deadlines are aggressive. The industry moves at a speed we hardly appreciate while we are in the middle of it.

A technical writer in the 1980s could have counted on six months or more to write a single manual, and was unlikely to have had to worry about formatting or graphics. Compare that to today's writer who might be very matter-of-fact about having a few weeks to write a similar manual. And that's not going to be the only thing on this writer's plate.—he is probably also working on a few other projects as well.

If you care about quality (don't we all?), it can be very discouraging to be perpetually unable to produce the documentation you know you're capable of and that the product deserves.

If you look at older industry standards for the amount of time it took to create technical documentation, you might feel jealous of those technical writers of yesteryear who expected to take seven to ten hours per page on documentation or close to that per help topic.

Those statistics were based on what were probably far fewer releases than you may be dealing with at your current job. When you have only one release every year or two, it's not hard to dedicate so much time to a single documentation deliverable. However, if you work in an environment that has many releases per year, add up the time you spend working on any given documentation project. You'll probably discover that you spent about seven to ten hours per page by the time you got it "right."

Documentation Always Comes Last

"Documentation always comes last" is a complaint of many writers, but it also happens to be something of a true statement. Documentation is rarely written until the product is developed into something that can be used, described, and recorded, so to some extent, it does come last in the process. (Even when a document is written early and based on specs, a good portion of it must be done toward the end of the release.)

When product managers and marketers plan a product, they often forget to tell the Technical Publications department. It's pretty frustrating to have an engineer ask nonchalantly if the documentation is finished yet for a product you have never even heard of, but unfortunately, it happens all the time.

Even when the writers are aware of the release schedule and willing to put in the effort needed to meet the deadlines, the developers might not be able to provide the right help, reviews, and information. Often, the developers work steadily and frantically up until the very last minute. They aren't always thinking about the tech writer, even if that writer has sent them several review copies, a bunch of email questions, and is thinking about ambushing them in the parking lot.

Because of the late start time for many documentation jobs, technical writers often have to work some very late nights at the end of a release.

> **TRUE STORIES**
>
> A tech writer can work long hours during an emergency. TJ remembers one tough period when he had to work with the product development team for a full month to do an extensive set of fixes for an Australian customer who wanted to see full documentation.
>
> This meant nearly constant work every evening and weekend. The work week started in California at 3:00 PM on Sundays, which was Monday morning in Australia. The only break was on Fridays when the Australian work week ended.
>
> TJ worked continually to document the changes as they were done. Although the work was hard and the hours long, there was satisfaction in being part of the team and a recognition of the importance of documentation from everyone in his company and the customer company.

Changes, Changes, Changes

If there's one thing that tech writers complain about the most, it's the fact that so many changes come in at the last minute. Because the tech writer does so much at the end of the release, he often has to deal with a shower of requests for changes at the last minute.

If you let yourself feel powerless and resentful because of it, you'll just stress yourself out. Remind yourself that everybody is a victim of the constant changes—not just you and your tech writing co-workers—and instead of taking it personally, just relax and accept it as part of the job. If you think of "doing major changes at the last minute, sometimes three or four times," as part of your job description, something you expect and can handle in a competent and professional way, you'll feel less like a victim.

Developers and the rest of the product team are suffering from last-minute changes, too; it's not just the tech writer. This whole "deadline" thing isn't much fun for anyone.

When You've Got to Pull a Rabbit Out of a Hat

So what do you do when you're asked to write a manual for a product that doesn't exist yet, has no specs and no prototype, and the one developer assigned to the project is too busy to speak to you?

When you're in a situation like this, don't be afraid. Look at it as an opportunity to gain some bragging rights. Find out what the product is going to do and how it will work. If it has a similar function to one that already exists, use the documentation for the other product as a basis for your new one and do your best to fill in the parts that you think will be needed. Once you have placeholders, you can fill in the information as it becomes available.

Afraid that your guesswork might be wrong? Write it anyway. It's a lot easier to remove or modify material in the eleventh hour than it is to start at the beginning and write it all from scratch at the last minute.

"I Don't Get No Respect"

If you feel that the tech writer is the lowly peon in the hierarchy, you aren't alone. It's a common complaint in many Tech Pubs departments, so much so that it became the cornerstone of Tina the Tech Writer's personality in *Dilbert*.

Is it true? Sometimes yes, sometimes no. Technical writers often don't get a lot of respect in high-tech companies. When a business is struggling to deliver a product, it often sees those with engineering backgrounds as the most important members of the company. Employees with "softer" skills or specialties in areas like technical writing or project management or others that are not directly related to getting a product out the door and in the hands of a customer are not always thought as highly of as we would like. And if you can't prove your worth or are seen as wasting the time of developers due to a lack of technical expertise, it can be a problem.

If you feel there's a lack of respect for your technical ability, take a good hard objective look at yourself and determine whether there are areas that can be improved. If there are, go do it with classes or self-study, and make sure you let the product team know what you're doing. In the long run, this technical expertise, along with the tips in this book, will help both you and your company.

If there is true disrespect for your role, figure out if it's embedded in the culture and not just the work of one sour individual. If it is greater than one person, the solution might be to get what you can out of the job, build your portfolio, and move on.

Working with Problem People

Yes, there are difficult people everywhere. We all have to deal with them on occasion. Why is this a special problem for the tech writer?

As a technical writer, you must depend on a lot of other people to help you do your job. Sometimes your job seems like one big coordination role as you pull together the contributions of product owners, designers, developers, and testers, many of whom understand only a small piece of the puzzle that makes up a product delivery. You sometimes have to nag these people to give you the information you need. And you have to maintain a pleasant working relationship with all of them, even though, to some of them, the relationship can seem as if they're doing all the giving and you're doing all the taking.

Sometimes it's just that they act as if you don't exist. They don't return your email, they don't look at your drafts. They avert their eyes when you pass them in the hall.

Others can behave in genuinely unpleasant ways. As a tech writer, you don't have the luxury of avoiding toxic people; you have to continue to deal with them. And you have to try to rise above the situation, because usually when two people have a bad relationship, others see it as the fault of both of them, not just one.

> **INSIDERS KNOW**
>
> If a co-worker's toxic behavior has legal repercussions for the company—that is, the offender is doing something that can be construed as sexual harassment or discriminatory in any way or is potentially dangerous—do not try to work things out between the two of you. Go to your manager immediately and make sure this reaches the attention of the Human Resources department.

Make sure that isn't the case with you. If you have a problem with one person, your manager is likely to be sympathetic and try to help. If you have problems with more than one person, people will start wondering what you're doing to cause the conflicts.

If someone attacks you in a way that is uncalled for, stand up for yourself and call him on his behavior, but don't stoop to his level. Fighting back in kind will only end up hurting you.

Then, pretend it never happened, and go right back in to work with the troublemaker. It can take a tough hide to march back into someone's cubicle after he has spoken to you rudely, but often, this is the best way to handle this kind of behavior. He could have been in a bad mood and have no idea how to control his feelings. He might be having personal problems that you don't know about. He might just be a jerk.

There are many good books out there for dealing with difficult people. Start with *Coping with Difficult People* by Robert Bramson and *Difficult Conversations: How to Discuss What Matters Most* by Douglas Stone.

Whether your difficult person is someone who won't give you the information you need, or is just plain mean, keep a "paper trail." Put everything in writing between yourself and the difficult person. Send this person detailed email explaining exactly what you are requesting and what you need. If a problem conversation occurs verbally, try to get a witness to the discussion and write notes as soon as the discussion is over. The paper trail can be helpful later if you need to explain why part of a job wasn't completed or why you feel you have to be transferred to a different project.

Remind yourself that it's not your fault if someone else has a problem. Sometimes the people with the worst behavior are brilliant workers, so companies overlook their personality defects. You can't change these people and you often can't choose not to work with them. You can only do your best.

Who Owns the Product?

If you're a person who loves to write and has pride in your profession, you might be a little disappointed by the realities of the tech writing life. Do you mind writing a book that won't have your name on it? Will you be unhappy when the next writer changes your elegant prose and rearranges the content you thought about so carefully?

Those are difficult hurdles for some people. At some point, you have to learn to think of your documentation as "product." And not everyone feels good about taking such an impersonal view of their work output.

As a paid employee or contractor of a company, you are producing work for hire. That means that even if you do all the work, the company that pays you owns your output. The company can give the work you did to any subsidiary company, customer, or other employee and let that recipient fold, spindle, or otherwise mutilate it.

No Upward Mobility

Technical writers don't always have a lot of upward mobility in their jobs. The Technical Publications department is relatively small, and besides managing that department, there aren't a lot of places for a technical writer to go.

Technical writing is usually seen as a specialty that does not necessarily apply to other parts of the business. In high-tech companies, Tech Pubs is often a small department within the Engineering division. It is extremely unlikely that a technical writer will be promoted to managing engineers, so often a writer can't see any future higher than a Tech Pubs manager.

The Two B's

A risk a tech writer runs is boredom on the job. You might find that surprising with so many different projects, so much stress, and so many last-minute changes going on, but it does happen. Working on the same kind of product and writing and maintaining the same documentation for years on end with little opportunity for advancement or job change can be boring.

This boredom, along with a sense of having little control over what goes on in the workplace, can make a tech writer a ripe candidate for job burnout. Burnout can cause a loss of interest in your work, inability to meet the demands of the department, and feelings of fatigue on all levels.

Many writers respond to this by moving regularly from one company to another, just for a change of pace, often finding themselves in the same boat there after a couple of years. Tech writing jobs have always been plentiful for writers who work in high-tech areas like Silicon Valley or New Jersey's pharmaceutical belt, but others don't have the option of moving on to a new job. The only effective way to avoid burnout is to acknowledge that it's a factor, prepare for it, deal with job stresses before they occur, and do what you can to take control of your work life.

Taking Control

You can't control what other people do, and often, at work, you feel you can't even control what *you're* expected to do. Nonetheless, there are ways to mitigate some of the madness, and it involves taking as much control as you can.

This book is all about ways to do that. Getting involved early, doing as much as you can in advance, and understanding the product well so you don't depend so much on developer input, all help.

Take the initiative. Don't wait around for a developer to give you information on the arcane aspects of the newest functionality in EnterPrize Cloud Storage version 3.0. Instead, write the content yourself. Take it to the developer and ask for confirmation of the areas you're not sure of. If your facts are wrong, your subject matter experts will correct them.

There are many problems you won't be able to solve no matter how much initiative and energy you have. Don't waste any more time and energy on those problems than you must. If you can't do anything about them, there comes a time when you must give up, accept the facts, and do what you can to take care of yourself, whether it's extreme exercise, massage, hot baths, or a much-needed vacation. You already know what you need to do, but you have to *do* these things, not just think about them or complain that you don't have the time.

Don't be put off by what I've told you in this chapter. I think it's important to go into any endeavor with your eyes wide open, and a bright tech writer does all the research necessary. In fact, the next chapter provides some helpful ideas about how to take charge of your career.

All in all, technical writing is not so dark after all. It's a satisfying job that offers you opportunities to learn, create, and deal with some very cool technology. Dark side? Nothing you can't deal with.

Chapter 24

Managing Your Career

Insights and guidance to help you advance in the right direction and still keep your options open.

What's in this chapter

- The different levels of technical writers
- Ways to move your career into more interesting directions
- Tips for how to keep your knowledge up to date
- Building and maintaining your people network
- Increasing your sphere of influence

We've almost reached the end of the book and you're just getting started.

The purpose of this book has been to get you through the technical writing door, and to help you get up to speed and be successful once you're inside. If you've followed the recommendations, you're well on your way.

But what happens after you've built some solid experience and you're no longer the neophyte with the wet feet? You need to continue to build on the career that this book has helped you start.

There comes a time in the lives of most technical writers when they'd like some sort of change. They might want more money or more challenge or more responsibility; they might want a different company and different assignments, or they might want to move in a completely different direction.

Moving Up

Cast your imagination into the future. You've been working as a tech writer for a few years now. You're doing good work and you enjoy your job, but you keep wondering one thing: *What's next?*

The answer depends on you. Many tech writers enjoy what they do so much that they don't feel a need to change. For many it's the perfect balance of structure and autonomy. You work independently most of the time and do a variety of different things every week, even every day. You interact with a lot of interesting and intelligent people and participate in exciting development work.

Other writers want more. As I discussed in the last chapter, there can be a serious lack of upward mobility in this field. What should you do when you have ambition and want to move ahead in your job?

For some, becoming a consultant or contractor satisfies that need. Chapter 22, "Working Outside the Box," talked about those options. Constantly changing venues and a higher hourly rate can provide job satisfaction to those who want to fly solo. But if that's not you, there are other paths to explore, such as looking for ways to move up or laterally within your company. And there's always the possibility of going to a different company altogether.

Before you look outside your current company, see what opportunities are available for you in your present job. Here are some technical writing levels that exist in many companies, with a brief description of what might be expected at that level.

Junior Technical Writer

A junior writer is an entry-level technical writer. At this level, employers expect you to be able to update and edit documents, proofread, and do some original writing under supervision. Nobody should expect a junior writer to write an entire document from scratch. Junior writers need, and should receive, learning time to develop expertise in the products, the tools, and the company's standards. After all, that's what "junior" is all about.

Intermediate Technical Writer

A mid-level tech writer can work with minimal supervision and could reasonably be expected to create a full document from scratch with little or no guidance. This writer knows how to follow a schedule and plans her work to meet deadlines. The intermediate tech writer is proficient in the tools needed to do the job, and has enough experience to quickly learn the tools and expectations of the job.

Senior Technical Writer

Senior writer is the top level for a tech writer in a nonmanagement position. (Some companies with large documentation departments have a higher level, a "principal" writer, as well.) A senior writer is a seasoned professional, able to work completely independently, and able to take a project from concept to finish with no guidance. Sometimes a senior tech writer is a lead writer, managing projects or leading and mentoring other less-experienced writers.

Documentation Manager

Not just any anyone can be a documentation manager. Documentation managers should understand the documentation development process from the ground up. And most of them do because most documentation managers have come up through the ranks.

If you're aiming for a management position, you need more skills than the ability to write. A manager's duties fall into four categories: planning, organizing resources, leading, and coordinating. On top of that, you'll probably be writing, too. Most managers still like to keep their hands in the game anyway, but it can be difficult to juggle your management and writing duties.

There are books to help you succeed as a documentation manager. Try *Managing Writers: A Real World Guide to Managing Technical Documentation* by Richard L. Hamilton, *Technical Writing Management: A Practical Guide* by Steven A. Schwarzman, or *Information Development: Managing Your Documentation Projects, Portfolio, and People* by JoAnn Hackos.

As the liaison between the Technical Publications department and the rest of the company, the manager has a big responsibility. The manager is also responsible for managing the budget and making sure Tech Pubs has the resources (both human and material) it needs to get things done. This means anticipating department needs based on the product roadmap, and getting requests through the proper channels so those needs are met on time.

Keeping Score

Whether you're a manager or an individual contributor, it's important to let upper management and the rest of the company know what your contributions are. Because Technical Publications is a *cost center* (a department that does not directly add to the profits of the company), it's all the more important for you to prove your worth.

Executives like to see metrics and scorecards, and technical writers don't always think in terms of showing off their accomplishments in this fashion. Think about what kinds of metrics you can put together to showcase the activities of the team. Whether it's number of deliverables, page count, on-time deliveries, or customer feedback, all of these can be used to show your worth to people who otherwise have no idea what technical writers do. Don't wait until review time to provide this information, either.

If you're wondering what these metrics might look like, think about what kind of information you can present to show that your output is progressing, or at least, that you are meeting requirements. This could be in the form of tables of information that provide monthly or quarterly information to upper management. You might want to present number of projects scheduled against projects delivered, or number of improvements made to existing documentation. You might want to show the number of on-time deliveries or number of unplanned projects completed. To create these presentations, you need some data to work with and some goals to work against.

Keeping Track

When managing a Tech Pubs department, I like to keep a spreadsheet of all work that comes in. This allows me to track completion and also to slice and dice by product family, document type, writers, due dates, and more. As each writer gets an assignment, the assignment goes into the spreadsheet and when it is completed, it's marked "Complete" in the spreadsheet. Because I track every job, I'm able to see at a glance what unexpected jobs come in, whether the writers deliver on time, whether a release date was moved, and whether one person has too much to do in a given period while another has too little.

Because technical writers do not want to be held responsible for delivering progressively more words each quarter (if we had to do that, everyone would be bloating their documentation to the point where it becomes unreadable), you can even set up a metric that shows number of excessive words eliminated in a given deliverable. This can be part of a quality goal.

Quality is a nebulous thing to track, but not impossible. First, consider the efforts you can make to improve quality in existing and upcoming documentation. This might be, as said above, reducing the number of words, creating targeted help content, creating customer-specific documentation, revising existing material, even applying templates for consistency.

Positive customer feedback is one of the best things you can present to upper management. While the quality improvements listed above are good examples of something you as a technical writer can plan to do, it's even better if you can ask customers what improvements they would like to see, and then make sure they get done. You can learn what customers want in various ways—through a survey, by telephone, or by sending direct email to the customers. You can even arrange to meet with them face to face.

Surveys are an excellent way to track customer satisfaction. There are several online survey companies, such as Survey Monkey at **surveymonkey.com** that offer low-cost ways to acquire user data.

Ask the customers specific questions about their opinions of the product documentation. Ask them if they've seen improvements over the past year and what those improvements might be. Tell them that you want to improve the product documentation and ask them what improvements they would like to see.

Don't limit the survey to yes/no questions. Make sure that many of your questions are open-ended so you can get some quotes to add to the reports you make. And ask for permission to follow up with anyone who fills out the survey. This way, you can ask more questions if you need to.

Plan to do a survey every year with the same questions. You'll have plenty of material for your yearly metrics, and plenty of material for your yearly goals.

Moving Around

Yes, a career in tech writing is not the corporate fast track to the CEO seat. The advantage, the benefit, even the appeal of a tech writing career is in the opportunities it offers for continuous learning, varied daily activities, and a high level of autonomy (all that and a good income, too).

Depending on your areas of interest and personal goals, you might not be sure what step to take after you reach senior-writer level. Not everyone wants to be a people manager. Frequently, senior employees assume a managerial role because there is no other place to go, even though they don't really want to be managers. When that happens, the company loses a good individual contributor and gains a less than fully engaged manager.

Moving Out

In tech writing, just as in other fields, sometimes you simply can't get what you want at your current job. If that's the case, the only solution is to go elsewhere.

The job-hunting advice in Chapter 4, "Breaking Into the Field," should be of some help in this endeavor. Whether it's your first job, second, or third, you still need to apply common sense, methodology, and focus to the hunt.

How Proprietary Is Proprietary Information?

When you start work at any company for which you produce work for hire, you sign a nondisclosure agreement (NDA). The NDA states, in essence, that you will not divulge company information.

Read the NDA before you sign it. (It's no excuse to later say you don't remember signing one.) Consider how it can affect you. If the documentation you write is proprietary, that means you are never allowed to show it to anyone else without permission.

That can be a problem when you are job-hunting and you want to show writing samples to the next potential employer. Some technical writers ignore the NDA and show the interviewer writing samples that are clearly marked "proprietary." Some interviewers don't object—or at least the writers think they don't, but then wonder why they aren't called back for the job.

Getting Portfolio Samples

The best time to resolve the portfolio question is before you sign the NDA, not when you're looking to get out of a job. Ask your hiring manager what documentation can be used in a portfolio. Some employers agree that you can show samples as long as you don't let them out of your sight during an interview or permit photocopying. Others allow you to show user guides or other documentation not deemed proprietary.

If your employer will not give permission to use any documentation samples, you can do what many writers do: disguise the documentation you wrote by eliminating company and product names and showing portions of it that don't contain proprietary information. At the interview, talk about the type of documentation it was, the target user, the process you went through to create it, and the problems and successes you had with it, and you should be fine.

Keep Your Knowledge Current

Your future is wide open, but you still have to make it through the intervening years. To make yourself more marketable in the short run and give yourself more options in the long run, always keep your knowledge and skills up to date. A busy tech writer doesn't always remember to plan for the future by thinking about what skills are important to learn.

For the most versatility, enhance your technical skills. (Sound familiar?) Take a class in programming or in the field you work in so you better understand the technology and customer needs (telecommunications, networking, semiconductors). Also consider classes in an area in which you might want to work. This can mean marketing, business, or project management.

Don't neglect your technical writing future, either. Try to keep up on what's new in technical writing. If you don't have experience with MadCap Flare or XML or DITA, look for ways to learn them so you can add them to your resume. Classes, seminars, and conferences are good places to learn and good places to make useful contacts.

Check into webinars and certification courses sponsored by the Society for Technical Communication (STC) at **stc.org/education**. Some employers will pay professional membership dues, conference costs, and webinar costs for you, so ask.

Moving into Another Field

Because people who make good technical writers also have qualities that make them good at other things, technical writers sometimes change careers after a while. There could be a place for you within your current company. Many departments need people who can write, have project management skills, and are familiar with the products. You might find an opportunity in Manufacturing, Marketing, User Experience, or Project Management, to name a few.

I know technical writers who have parlayed their experience and expertise into jobs as system administrators, sales engineers, product managers, project managers, user researchers, trainers, programmers, linguistic researchers, interaction designers, and recruiters. Pretty much everything you do as a technical writer can be a doorway into your second—or third or fourth—career.

Networking Doesn't Stop Just Because You're Employed

Network, network, network! I said it in Chapter 4, "Breaking Into the Field," and I can't overemphasize the importance of doing so. You might be happy at your job now, but we all know that things can suddenly happen to jobs that seemed secure only yesterday. You should continue to make contacts during your entire working life.

And networking doesn't just mean meeting other writers. It means meeting anyone who works in a field you, or one of your acquaintances, might want to work in. Networking goes both ways. If you can help someone else fill a job or obtain one, do it! The people involved will remember you favorably later when you need assistance yourself.

Even if you never leave your present position (and how likely is that?), networking is beneficial. Knowing other technical writers helps when your own Tech Pubs department has an opening, for example. And mixing with other tech writers can help you learn about and get recommendations for solutions to work problems. If you can call on a pool of experienced writers, you can save yourself critical time and effort in a crisis.

TRUE STORIES

Meryl's story: My manager encouraged me to speak or publish outside of my company. I was afraid of public speaking, and not sure I had anything interesting to publish, but took a chance and submitted a proposal to speak at a conference. My proposal was accepted, and I was invited to speak! I started working furiously to develop my ideas and joined Toastmasters to overcome my fear of public speaking.

Though I was nervous, the conference presentation went over well and people liked my ideas. One attendee had traveled across the country because the description of my talk had piqued her interest. Since then, I have been invited to speak at other conferences and become active in Toastmasters. I have even been paid to speak.

The main lesson here for me is that I really do have good ideas, so why not share them with others and get some recognition for my efforts?

The recommendations in Chapter 4, "Breaking Into the Field," for joining a community of technical communicators still hold true at any stage in your career. Join the Society for Technical Communication (STC) at **stc.org** and attend local meetings. It can be even more fun and interesting if you volunteer for the organization.

The TECHWR-L mailing list and its associated website at **techwhirl.com** will introduce you to a worldwide community of fellow technical communicators. LinkedIn has many user groups for technical communicators, and you can start or participate in discussions for each group you join.

Know Your Peers

There are many conferences of interest to tech writers in North America and Europe, sponsored by organizations that you should know about. Check out the websites below not only for conference information.

- TechComm Summit sponsored by STC

  summit.stc.org

- The Software User Assistance Conference sponsored by Writers UA

  writersua.com

- The European UA conference sponsored by UA Europe

uaconference.eu

- Technical Communication UK conference

  technicalcommunicationuk.com

- LavaCon Conference on digital media and content strategies

  lavacon.org

- Usability Professionals' Association (UPA) international conference

  usabilityprofessionals.org

- Intelligent Content Conference sponsored by the Rockley Group

  rockley.com

- Tekom Conference for technical communication and information development (Germany)

  tekom.de

- Gilbane Content Management Conference

  gilbane.com

- Content Management Strategies /DITA North America Conference hosted by the Center for Information-Management Development

  cm-strategies.com

- Best Practices Conference

  infomanagementcenter.com

"On the Internet, Nobody Knows You're a Dog"

This classic line about what's possible on the Internet still applies.

In today's online world, anyone can create a website. Why not create one to let people see how good you really are? Having a website not only provides you with an opportunity to post your resume where unlimited numbers of people can see it, but also gives you a forum to share what you've learned as a technical writer. If you have samples you're allowed to show, you can post them here and point prospective employers to the site.

Share the Wealth

As you progress in your career and develop your contacts, you might come to feel that you have something more to offer. Instead of simply attending conferences, you might have ideas for presenting at conferences. Instead of just reading blogs, you could write a blog of your own.

Paying it forward can end up helping you in more ways than you know.

Think about what experience you have that is worth sharing. Presenting at conferences or STC meetings is an excellent way to increase your visibility in the technical writing world and thus increase your marketability as well. Look at some of the conference websites to see what kinds of sessions they are scheduling and review their criteria for submitting presentations. You're sure to have some ideas for a presentation of your own.

Toastmasters International is a world leader in communication and leadership development. If you've ever felt that you could use help in developing your speaking skills, Toastmasters might be just what you need. If you enjoy it, consider starting a Toastmasters club at your company. Go to **toastmasters.org** to learn more.

You might also look for ways to contribute to local chapters of some of the professional organizations mentioned throughout *The Insider's Guide* by volunteering or running for office.

Do something to help others who are just getting started, like you were once, by teaching or mentoring. Community colleges and technical writing certificate schools can offer opportunities to teach what you live every day. It can be a satisfying experience, offer variety in your work, and often provide a bit of extra income. And it feels good to give back.

The Future Belongs to You

Although your life as a technical writer might be so rewarding that you never feel the need to make a major career change, it's nice to know that you can. Start preparing now for your future by exploring all the paths that are open to you. The things you do today to help yourself today are what will make your working life better tomorrow—and for years to come.

Appendixes

Appendix A

Tech Talk: The Tech Writer's Glossary

Definitions for some of the words and terms used in this book.

---

active voice. The voice used to indicate that the subject of a sentence is directly performing the action expressed by the verb.

Agile. A software development methodology based on iterative and incremental development.

Ajax. (Asynchronous JavaScript and XML) A method of combining interactive applications such as JavaScript, dynamic HTML, XML, CSS, and others on a web page.

API. (Application Programming Interface) An interface that enables one program to interact with another.

application. Software that lets a user perform a particular task or set of tasks.

ascenders. The highest part of a character in a typeface.

ASCII. ASCII (American Standard Code for Information Interchange) is one of the oldest methods for encoding characters for use in computers. It contains 128 characters, including the English alphabet, punctuation, and other characters. Most text processing programs no longer use ASCII, instead they use Unicode, which includes the ASCII characters, but can also handle nearly all human languages. See Unicode.

assistive technology. Software or devices designed to be used by people with disabilities.

backlog. Prioritized list of requirements used in the Agile development process.

baseline. The line upon which the letters of a typeface "sit." The bottom of an "x" is typically on the baseline.

best practices. A method or technique that has consistently shown superior results and is often used as a benchmark.

bitmap. An image file, sometimes called a *raster* file, that is made of pixels. *See* raster file.

build (software). The process of compiling source code to turn it into finished software that runs on a computer.

burnout. A state of physical, emotional, and mental exhaustion, sometimes marked by apathy or resentment, usually caused by long-term work situations that are stressful and demanding.

C, C++. Programming languages.

callout. A caption that contains a pointer to an area of interest.

change bar. A vertical bar in the margin of a document that indicates where content is different.

Cascading Style Sheet. *See* CSS.

conditional text. Text within a document that is intended to appear in some versions of the document, but not others.

consultant. An independent worker who is hired by a company to solve a particular problem.

content reuse. The management of content by breaking it into small enough components, or topics, so that each topic can be used in the appropriate place.

context-sensitive. Directly related to the nearby content. When describing help, this means that the help relates to a specific area of a web page. *See also* page-sensitive.

contractor. A temporary worker who is hired to work on a specific project or set of projects, for a specified amount of time.

copyright. The exclusive ownership, protected by law, of a literary work or other work of art. The copyright confers the right to make use of the work.

cross-functional. Consisting of individuals from more than one organizational unit or function.

cross-platform. Used on different operating systems.

CSS. Cascading Style Sheet, a language for specifying how a web page is presented.

descenders. The lowest part of a character in a typeface.

design document. Document that describes how the product works, and describes the product architecture.

development process. The process a company goes through to take its product from concept to finished work. *See also* SDLC.

DITA. (Darwin Information Typing Architecture) An XML schema for authoring and publishing topic-oriented content.

DocBook. An XML schema for authoring and publishing technical documentation.

documentation. any content (written, illustrated, or both) supporting the use, operation, maintenance, or design of a product or service.

documentation plan. A specification that describes what a document or set of documentation will consist of and what its schedule is.

domain expert. Subject matter expert in a specific field or endeavor. In high tech, this generally refers to knowledge in a field other than software.

draft. Any iteration of a document before it is finalized.

DTD. (Document Type Definition) A schema language used to define the elements, attributes, and structure of an XML-based language.

ebook. An electronic book, readable on a computer or other electronic device.

end point. The state or condition that happens when a procedure reaches its natural conclusion.

end user. The person who uses a product or service (as opposed to the person who buys it, manages it, or designs it).

enterprise software. Software used in business or government as opposed to that used by individuals.

escalation path. The escalating steps a customer takes when trouble occurs, starting with a proposed solution for the problem, and ending with a call to the company for assistance if the solutions don't work.

external facing. Materials that are meant to be seen by customers and others outside the organization.

extranet. A portion of an organization's internal network that is accessible to a controlled set of outside users, such as customers or vendors, but not to the general public.

FAQ. (Frequently Asked Questions) A set of basic questions and their answers about a specific topic or product.

flush. Aligned, as in type.

font. A single size and style of a particular typeface, although in the digital world, "font" is typically used to refer to any typeface. *See* typeface.

freeze. The point in development where all activity stops.

gerund. A verb form that acts as a noun, ending in "-ing."

glossary (translation). A company-specific list of words and their accepted translation into different languages.

grid. An invisible structure of horizontal and vertical lines that defines where elements go on a web page or document page.

hard copy. Documentation printed on paper; the opposite of electronic, or "soft copy."

help authoring tool (HAT). Program used to write and generate online help

heuristic evaluation. Expert review of a user interface or product against standard usability best practices

HLDD (High-Level Design Document). A document that describes the software architecture and how the software works. *See also* LLDD.

hover. Movement of the mouse pointer over an icon or link, without clicking a button, that causes a change to the application. *See* mouseover.

HTML. Hypertext Markup Language, the code that uses tags to tell web browsers how to display content.

human factors. A discipline of study, sometimes called ergonomics, that involves the study of the way humans react with their environment.

imperative mood. The form of a verb that makes a direct command.

infinitive. Verb form that shows no person, or tense. Usually the "to" form of the verb, although the imperative mood also uses the infinitive without "to." *See also* imperative mood.

informational interview. An interview that is conducted by a would-be employee for the purpose of collecting information rather than seeking a job.

internal facing. Materials that are meant to be seen only by people within the organization.

IT. Information Technology (pronounced "Eye-Tee").

Java. a programming language designed for Internet development.

JavaScript. A programming language that is used mainly to create dynamic, interactive web pages.

JIRA. An issue-tracking system.

justified type. Type that is aligned straight on both the left and right sides.

knowledge base. A centralized repository of information, from which internal and external users seek help.

landing page. *See* splash page.

layers. In image editing, additional drawing areas that can be overlaid on one another to add greater control. A layer can be used to add text to an image. Layers can also be hidden or shown to allow a single image file to be generated in many different ways.

leading. The vertical space between lines of text.

Linux. (Pronounced "Linnucks") A UNIX-like operating system designed to provide UNIX capabilities at low cost or free.

LLDD (Low-Level Design Document). A document that describes the software architecture and how the software works in more detail than a HLDD. *See also* HLDD.

localization. The act of changing content so that it is relevant to the locale.

locator (in indexing). The part of the index entry that takes the user to the target. It can be a page number, a range of page numbers, or a pointer to a different index entry. Also called reference. *See also* reference.

lorem ipsum. Dummy content made up of Latin text, used by designers to indicate text in a layout.

mailing list. An email discussion group.

marcomm. Marketing Communications, the messages and media used to interact with customers.

marker (in indexing). An indicator that the marked word or term belongs in the index.

metrics. A set of measurements by which performance can be assessed.

milestone. An important action or event on a timeline.

monospaced font. A font in which all characters use up the same amount of space. A lowercase "i" takes the same amount of space as a capital "M."

mouseover. Movement of the mouse pointer over an icon or link, without clicking a button, that causes a change to the application. *See* hover.

MRD (Marketing Requirements Document). A document that describes what a product must include to meet customer needs.

NOC (Network Operations Center). Data center where a company's servers and networking equipment are located.

online help. Task-oriented modules (sometimes called assistive documentation) that come up when a user clicks a Help link or icon on a software application or web page.

open source software. Software whose source code is openly shared with and among developers and users.

outdent. *A* line that extends outside of the normal margin.

out-of-the-box experience . (OOBE) The impression a user has when first opening a package and setting up the product.

page-sensitive. Used to refer to help that relates to the information on a single page. *See also* context-sensitive.

passive voice. The voice used to indicate that the subject of a sentence is the recipient of the action expressed by the verb.

PDF. (Portable Document Format) A file format originally created by Adobe Systems that provides an electronic image of text and graphics that looks like a printed document.

peer editing. Editing done by a colleague of equal standing.

person. In writing, a way to indicate how near the reader or writer is to what is being said. The first person is "I," "me," or "we." The second person is "you." The third person is "he," "she," "it," "they," or "them."

personas. Fictional characters designed to represent the target users of a given product.

placeholder. A substitute piece of text or graphic used temporarily in place of the real thing.

point. A measurement that equals 1/72 of an inch, used to measure type sizes.

PRD (Product Requirements Document). A document that defines a product and the features it must have.

procedure. Numbered steps that guide a user through a process.

product manager. The person responsible for selecting and determining the features of a product and overseeing it as it goes through development.

proofreading. The process of reviewing and correcting content for typos, grammar errors, and stylistic issues.

proprietary. Belonging to someone or something that has exclusive rights of ownership.

QA (Quality Assurance). A system process of testing to determine that a product or service meets specifications.

quick start. A short document that is designed to get the user immediately up and running.

ragged. A typography term meaning there is no straight alignment. The text in this book is flush left, ragged right.

raster file. An image file, sometimes called a *bitmap*, that is made of pixels. *See* bitmap.

README. A file that accompanies software and is intended to be read first.

redundant pairs. Words that people tend to put together out of habit but one word is unneeded.

reference (in indexing). Also called locator, the part of the index entry that takes the user to the target. It can be a page number, a range of page numbers, or a pointer to a different index entry. *See also* locator.

requirements. The features and functionality that are needed in a particular release of a product.

roadmap. In high-tech, a product development plan that matches goals with technology solutions.

run book. Set of written procedures for operation of the system or network by an administrator or operator.

sandbox. An environment in which a person can try out software to see how it works.

sans serif. Typeface without serifs. (*See* serif.)

screen shot. An image recording of a portion of the information on the computer screen.

scrum. An Agile development method based on defined development periods called *sprints*. *See also* Agile. *See also* sprints.

SDLC (Software Development Life Cycle). Project management model that describes the software development stages. *See also* development process.

search engine optimization (SEO). The process of improving a website's visibility so more visitors come to it.

second person. Designation of the person to whom you are speaking as the subject: "You."

Section 508. An amendment to the US Rehabilitation Act to require federal agencies to make electronic and information technology accessible to people with disabilities.

self-help. A customer's ability to access knowledge base to find information, thus avoiding calls to Customer Support.

sentence case. A way of capitalizing that uses capital letters for the first word and all proper nouns in the segment.

serial comma. A comma that precedes the final item in a list. (Also known as the Oxford comma.)

serif. Short cross-line at the end of the main strokes of a letter in a typeface like Times, Palatino, or Georgia.

server. A computer that manages network resources.

Simplified Technical English. A controlled language used primarily in aerospace maintenance documentation.

single-sourcing. Developing content with the intent of producing all or parts of it, in multiple formats.

SME. *See* subject matter expert.

social media. The use of web-based and mobile technology to create interactive dialog.

specification. Document that defines what a product or application does.

splash page. Initial web page (sometimes called landing page) that acts as an introduction to the rest of the site.

sprint. In the Agile scrum development method, a unit between a week and a month during which the development team works on a defined set of features and functionality.

SQL. Structured Query Language. A programming language used to manage or access data in a relational database.

stakeholder. A person that has an interest in a specific project or business activity.

standup meeting. Short, usually daily, status meeting with members of an Agile scrum team.

stet. A proofreading mark that means "Don't change this."

structured content. Content that has been classified using metadata (data that describes other data).

structured authoring. Writing that follows the enforcement of organizational structure of content components.

subject (in indexing). The word, phrase, or abbreviation listed in alphabetical order in the index. Also called topic. *See also* topic.

subject matter expert (SME). The person who knows about a specific aspect of technology. *See also* domain expert.

surrogate user. In usability testing, a participant who has the characteristics and business needs of the customer to whom the product is targeted.

technical writer. Someone who conveys information about a technical subject, directed at a specific audience for a specific purpose.

template. A preformatted file that is used to create other documents.

tense. In grammar, the time of action (present, past, or future).

third person. Designation of a person other than yourself or the one you are speaking to, with pronouns like *he, she, they,* or *them.*

time to market. The length of time from the beginning of product development to its availability for customer delivery.

title case. A way of capitalizing that applies capital letters to the first and last words of a title and all nouns, pronouns, adjectives, verbs, adverbs, and subordinating conjunctions. *See also* sentence case.

topic (in indexing). The word, phrase, or abbreviation listed in alphabetical order in the index. Also called subject. *See also* subject.

topic-oriented writing. Writing intended for reuse. Each topic is a unit of information that stands alone, or can be mixed and match with other units. *See also* content reuse.

translation memory. a database of text segments that have been translated.

trigger. In a procedure, the event that starts procedure.

typeface. A single set of characters..

troubleshooting. The act of investigating the cause of problems.

Unicode. A standard for encoding text for use in computers. Unicode supports most of the world's writing systems. *See* ASCII. *See* UTF-8.

UNIX. An operating system widely used in workstations and servers.

usability. The ease of use with which a human being interacts with a product; the act of taking physical and psychological requirements of human beings into account during the design process.

use case. A description of a system's behavior in response to user actions.

user-centered design. Iterative design tested with surrogate customers until the majority of test subjects succeed at completing the intended tasks.

user experience. (UX) The entire interaction a person has with a product, system, or service. User Experience professionals focus on improving the usability of a product.

user-friendly. Easy to learn and use.

user interface. (UI) The set of commands through which a human being interacts with a computer; typically, a graphical interface with windows, icons, and menus.

user stories. In a scrum process, short statements about what the user does with the product. *See also* scrum.

UTF-8. A mapping method for representing Unicode characters that is widely used in text processing and web applications. UTF-8 is compatible with ASCII. See ASCII. See Unicode.

UX. *See* user experience

vector file. A graphic file made of lines and points, which can be enlarged and reduced without losing data.

voice. A grammatical term that describes how the subject and verb in a sentence related to each other. *See* active voice *and* passive voice.

Waterfall. A software development process in which development is done sequentially.

web server. A computer that "serves up" web content to a browser.

web services. Machine-to-machine interaction over a network

WFH. Acronym for "working from home."

white paper. A document (like a report) that states a position or helps to solve a problem. White papers are often written in a style that is part way between marketing language and technical language.

white space. The negative space on a page.

wiki. A collaborative website that can be modified quickly by many users.

workaround. A solution that enables a user to "work around" a problem rather than fixing the problem.

workflow. A set of tasks and steps that make up a work process.

work for hire. A work created by one person and paid for (and thus owned) by another (the employer).

x-height. The height of a basic lowercase letter in a typeface, such as the height of an "x."

XML. (eXtensible Markup Language) A metalanguage (language used to describe or analyze language) that is used to define other languages. *See* DITA. *See* DocBook.

XML schema. A formal description of what an XML document can contain.

Appendix B

For Your Bookshelf

A list, far from exhaustive, of books you might find useful. Some have been around for years but are still good sources of information on their topic.

Style Guides

The Chicago Manual of Style
By University of Chicago Press Staff
University of Chicago Press; 16th edition

The Columbia Guide to Online Style
By Janice R. Walker and Todd Taylor
Columbia University Press; 2nd edition

The Elements of Style
By William Strunk and E. B. White
Longman; 4th edition

The Global English Style Guide: Writing Clear, Translatable Documentation for a Global Market
By John Kohl
SAS Press

The IBM Style Guide: Conventions for Writers and Editors
By Francis DeRespinis, Peter Hayward, Jana Jenkins, Amy Laird, Leslie McDonald, Eric Radzinski
IBM Press

Microsoft Manual of Style
By Microsoft Corporation
Microsoft Press; 4th Edition

Read Me First! A Style Guide for the Computer Industry
By Sun Technical Publications
Prentice Hall; 3rd edition

The Yahoo! Style Guide: The Ultimate Sourcebook for Writing, Editing, and Creating Content for the Digital World
By Yahoo!
St. Martin's Griffin

Writing Resources

The Blue Book of Grammar and Punctuation
By Jane Strauss
Jossey-Bass; 10th edition

Developing Quality Technical Information: A Handbook for Writers and Editors
By Gretchen Hargin, Michelle Carey, Ann Kilty Hernandez, Polly Hughes, Deirdre Longo, Shannon Rouiller, Elizabeth Wilde
IBM Press; 2nd edition

Handbook for Writing Proposals, Second Edition
By Robert J. Hamper and L. Sue Baugh
McGraw-Hill; 2nd edition

Handbook of Technical Writing
By Gerald J. Alred, Charles T. Brusaw , Walter E. Oliu
St. Martin's Press; 10th edition

How to Communicate Technical Information: A Handbook of Software and Hardware Documentation
By Jonathan Price and Henry Korman
Addison-Wesley Professional

Kaplan Technical Writing: A Comprehensive Resource for Technical Writers at All Levels
By Carrie Hannigan, Carrie Wells, Diane Martinez, Carolyn Stevenson, and Tanya Peterson
Kaplan Publishing; 2nd edition

Letting Go of the Words: Writing Web Content that Works
By Janice (Ginny) Redish
Morgan Kaufmann

Spring Into Technical Writing for Engineers and Scientists
By Barry J. Rosenberg
Addison-Wesley Professional

Technical Writing 101: A Real-World Guide to Planning and Writing Technical Content
By Alan S. Pringle and Sarah S. O'Keefe
Scriptorium Publishing Services; 3rd edition

Tools and Technology

Beginning CSS: Cascading Style Sheets for Web Design
By Ian Pouncey and Richard York
Wrox; 3rd edition

Building a WordPress Blog People Want to Read
By Scott McNulty
Peachpit Press; 2nd edition

The Compass: Essential Reading about XML, DITA, and Web 2.0
By Sarah S. O'Keefe
Scriptorium Publishing; 2nd edition

Conversation and Community: The Social Web for Documentation
By Anne Gentle
XML Press

CSS: The Missing Manual
By David Sawyer McFarland
O'Reilly Media; 2nd edition

DITA Best Practices: A Roadmap for Writing, Editing, and Architecting in DITA
By Laura Bellamy, Michelle Carey, and Jenifer Schlotfeldt
IBM Press

Introduction to DITA: A User Guide to the Darwin Information Typing Architecture Including DITA 1.2

By JoAnn T. Hackos

Comtech Services, Inc.; 2nd edition

XML in a Nutshell

By Elliotte Rusty Harold, W. Scott Means

O'Reilly Media 3rd edition

XML in Technical Communication

By Charles Cowan

Institute of Scientific and Technical Co

WIKI: Grow Your Own for Fun and Profit

By Alan J. Porter

XML Press

Content Design and Planning

Content Management for Dynamic Web Delivery

By JoAnn T. Hackos

Wiley

Content Strategy for the Web

By Kristina Halvorson and Melissa Rach

New Riders Press; 2nd edition

Don't Make Me Think: A Common Sense Approach to Web Usability

By Steve Krug

New Riders Press; 2nd edition

Is the Help Helpful? How to Create Online Help That Meets Your Users' Needs

By Jean Hollis Weber

Hentzenwerke Publishing

A Practical Guide to Localization

By Bert Esselink

John Benjamins Publishing Co.

User and Task Analysis for Interface Design
By JoAnn T. Hackos and Janice C. Redish
Wiley

Web Style Guide: Basic Design Principles for Creating Web Sites
By Patrick J. Lynch and Sara Horton
Yale University Press; 3rd edition

People and Project Management

Information Development: Managing Your Documentation Projects, Portfolio, and People
By JoAnn T. Hackos
Wiley; 2nd edition

Managing Your Documentation Projects
By JoAnn T. Hackos
Wiley

Managing Translation Services
By Geoffrey Samuelsson-Brown
Multilingual Matters

Managing Writers: A Real World Guide to Managing Technical Documentation
By Richard L. Hamilton
XML Press

Technical Writing Management: A Practical Guide
By Steven A. Schwarzman
CreateSpace

And More...

The 4-Hour Workweek: Escape 9-5, Live Anywhere, and Join the New Rich
By Timothy Ferriss
Crown Archetype

Coping with Difficult People: The Proven-Effective Battle Plan That Has Helped Millions Deal with the Troublemakers in Their Lives at Home and at Work
By Robert M. Bramson
Dell/Random House

Difficult Conversations: How to Discuss What Matters Most
By Douglas Stone, Bruce Patton, and Sheila Heen
Penguin

Demystifying Grant Seeking: What You Really Need to Do to Get Grants
By Larissa Golden Brown and Martin John Brown
Jossey-Bass

How to Measure Anything: Finding the Value of Intangibles in Business
By Douglas W. Hubbard
Wiley

The Inmates Are Running the Asylum: Why High-Tech Products Drive Us Crazy and How to Restore the Sanity
By Alan Cooper
Sams Pearson Education

The Non-Designer's Design Book
By Robin Williams
Peachpit Press; 3rd edition

Succeeding with Agile: Software Development Using Scrum
Mike Cohn
Addison-Wesley Professional

Writing Effective Use Cases
Alistair Cockburn
Addison-Wesley Professional

Appendix C

Websites

The websites listed in this book, plus a few more that you'll like. Websites come and go, but all were live at the time of publication.

Writing Resources

The Content Wrangler
Content is a business asset worthy of being managed
http://thecontentwrangler.com

Doc Symmetry
Technical writing tips for technical writers and managers of writers
www.docsymmetry.com

English Grammar
Grammar lessons, exercises, and rules for everyday use
www.englishgrammar.org

Grammar Book
Website for *The Blue Book of Grammar and Punctuation*
www.grammarbook.com/

Grammar Girl
Quick and dirty tips for better writing
grammar.quickanddirtytips.com

Gryphon Mountain Journals
Ben Minson's blog on technical communication
www.gryphonmountain.net

Guide to grammar and writing
http://grammar.ccc.commnet.edu/grammar

Plain Language
Improving communication from the federal government to the public
www.plainlanguage.gov

The Rockley Group
Content management strategies
www.rockley.com

Simplified Technical English
International specification for preparing documentation in a controlled language
www.asd-ste100.org

Writing.com
An online community for writers
www.writing.com

Structured Content

XML
www.xml.com
www.w3.org/XML

DITA
www.ibm.com/developerworks/xml/library/x-dita1
www.ditausers.org
dita.xml.org

DocBook
www.docbook.org

Sites to Help Build Your Portfolio

Calibre
ebook management
www.calibre-ebook.com

FLOSS Manuals Foundation
Free manuals for free software
www.flossmanuals.org

Grant Writers Online
Your guide to easy and effective grant writing
grantwritersonline.com

iFixit
The free online repair manual that you can edit
www.ifixit.com

Open Office.org
The Free and Open Productivity Suite
www.openoffice.org

Open Source as Alternative
Open source software alternatives
www.osalt.com

Open Source Initiative
Education on open source
www.opensource.org

Open Source Windows
Open source software that runs on Windows
www.opensourcewindows.org

Sourceforge
Find, create, and publish open source software for free
www.sourceforge.net

Technical Writing Zone
Helping you find writing jobs
www.technicalwritingzone.com

Volunteer Match
volunteermatch.org

wikiHow
The how-to manual you can edit
www.wikihow.com

Wikipedia
The free, editable online encyclopedia
www.wikipedia.com

Networking and Job-Hunting

Ask the Headhunter
Nick Corcodilos's website on job searching and hiring
www.asktheheadhunter.com

Ask the Headhunter blog
www.corcodilos.com/blog

LinkedIn
www.linkedin.com

Meetup
www.meetup.com

Technical Writing World
The social network for technical communicators
http://technicalwritingworld.com

United States Department of Labor Bureau of Labor Statistics
Technical Writer Job Category
bls.gov/ooh/Media-and-Communication/Technical-writers.htm

Organizations and Conferences

Center for Information-Development Management
www.infomanagementcenter.com

Center for Information-Management Development
www.cm-strategies.com

LavaCon Conference on Digital Media and Content Strategies
http://lavacon.org

Project Management Institute
www.pmi.org

Rockley Group
rockley.com

Society for Technical Communication (STC)
www.stc.org

Techwhirl, home of the TECHWR-L mailing list
www.techwhirl.com

Technical Communication UK
www.technicalcommunicationuk.com

Tekom (German association for technical communication and information development)
www.tekom.de

Toastmasters International
www.toastmasters.org

Usability Professionals' Association (UPA)
www.usabilityprofessionals.org

Writers UA
Training and information for user assistance professionals
www.writersua.com

Technical Dictionaries

Acronym Finder
www.acronymfinder.com

Tech Dictionary
www.techdictionary.com

TechTerms
www.techterms.com

The Lighter Side of Technical Writing

Ban Comic Sans
Putting the *sans* in Comic Sans
www.bancomicsans.com

Comic Sans Criminal
"Helping people like you learn to use Comic Sans appropriately"
www.comicsanscriminal.com

Dilbert
www.dilbert.com

Miscellaneous

Blind Text Generator (Lorem Ipsum generator)
www.blindtextgenerator.com/lorem-ipsum

eBay
www.ebay.com

JIRA Issue-Tracking Tool
atlassian.com/JIRA

Keirsey Temperament Sorter
www.keirsey.com

Myers-Briggs
www.myersbriggs.org

Readability Formulas
www.readabilityformulas.com

Safe Computing Tips
www.safecomputingtips.com

Telework Research Network
www.teleworkresearchnetwork.com

User Interface Engineering
Jared Spool's firm specializing in website and product usability
www.uie.com

WebMD Stretching Exercises at Your Desk
www.webmd.com/fitness-exercise/features/stretching-exercises-at-your-desk-12-simple-tips

WikiHow - How to Exercise While Sitting at Your Computer
www.wikihow.com/Exercise-While-Sitting-at-Your-Computer

Index

C

D

Q

R

S

U

V

W

X

Lightning Source UK Ltd.
Milton Keynes UK
UKHW030634070820
367857UK00008B/1108

9 781937 434038